Numerical Analysis:
An Introduction

Numerical Analysis: An Introduction

Griffin Cook

WILLFORD PRESS

www.willfordpress.com

Published by Willford Press,
118-35 Queens Blvd., Suite 400,
Forest Hills, NY 11375, USA

ISBN: 978-1-68285-723-6

Cataloging-in-Publication Data

Numerical analysis : an introduction / Griffin Cook.
p. cm.
Includes bibliographical references and index.
ISBN 978-1-68285-723-6
1. Numerical analysis. 2. Mathematical analysis. 3. Mathematics. I. Cook, Griffin.
QA297 .N86 2019
519.6--dc23

For information on all Willford Press publications
visit our website at www.willfordpress.com

WILLFORD PRESS

Contents

Permissions

Index

Contents

Permissions

Index

Preface

The study of algorithms, which use numerical approximation for the solution of problems of mathematical analysis, is termed as numerical analysis. It has applications in the diverse areas of physical sciences and engineering. Direct and iterative methods and discretization are generally employed in numerical analysis. An important area of this field is the study of errors. Errors are introduced in the solutions obtained through numerical analysis due to round-off, truncation and discretization methods. The major sub-disciplines within this field are interpolation, extrapolation and regression, mathematical optimization, numerical integration, numerical differential equation, etc. This textbook provides comprehensive insights into numerical analysis. Most of the topics introduced in this book cover the principles and methodologies used in numerical analysis. This textbook, with its detailed analyses and data, will prove immensely beneficial to professionals and students involved in this area at various levels.

A short introduction to every chapter is written below to provide an overview of the content of the book:

Chapter 1- The study of algorithms, which use numerical approximation for the solution of problems in mathematical analysis, is called numerical analysis. This is an introductory chapter, which will introduce briefly all the significant aspects of numerical analysis, its principles and use;

Chapter 2- The study of algorithms aimed at the operation of linear algebra computations on computers is under the scope of numerical linear algebra. It has applications in image and signal processing, computational finance, telecommunication, materials science simulations, etc. This chapter has been carefully written to provide an easy understanding of numerical linear algebra through a detailed discussion of systems of linear equations, Gauss elimination method, pivoting, LU decomposition, Cholesky decomposition, etc.

Chapter 3- In science and engineering, it is common to obtain a set of data points through sampling or experimentation. These represent the values of a function corresponding to values of an independent variable. To estimate the value of the function at on intermediate value of the variable, it is necessary to construct new data points in a process called interpolation. The aim of this chapter is to explore the fundamentals of interpolation, which includes the topics, polynomial interpolation, newton divided difference, LaGrange polynomial and cubic spline interpolation.

Chapter 4- Numerical integration comprises a family of algorithms that are meant for the calculation of the numerical value of a definite integral. It also encompasses the solution of differential equations using numerical methods. The topics elaborated in this chapter will help in developing a better perspective about truncation error, mid point rule, trapezoidal rule, Simpson's rule, Newton–Cotes formulas and Gaussian quadrature, for a comprehensive understanding of numerical integration.

Chapter 5- It is not possible to find analytical solutions to many differential equations. Often, a numeric approximation to the solution will suffice. Various numerical methods can be used for obtaining such approximations. This chapter closely examines the key concepts of numerical differential equations and elucidates the finite difference method, numerical methods for ordinary differential equations and general linear methods.

Finally, I would like to thank my fellow scholars who gave constructive feedback and my family members who supported me at every step.

Griffin Cook

Introduction to Numerical Analysis

The study of algorithms, which use numerical approximation for the solution of problems in mathematical analysis, is called numerical analysis. This is an introductory chapter, which will introduce briefly all the significant aspects of numerical analysis, its principles and use.

In the area of mathematics and computer science, Numerical analysis creates, analyzes, and implements algorithms for obtaining numerical solutions to problems involving continuous variables. Such problems arise throughout the natural sciences, social sciences, engineering, medicine, and business. Since the mid-20th century, the growth in power and availability of digital computers has led to an increasing use of realistic mathematical models in science and engineering, and numerical analysis of increasing sophistication is needed to solve these more detailed models of the world. The formal academic area of numerical analysis ranges from quite theoretical mathematical studies to computer science issues.

With the increasing availability of computers, the new discipline of scientific computing, or computational science, emerged during the 1980s and 1990s. The discipline combines numerical analysis, symbolic mathematical computations, computer graphics, and other areas of computer science to make it easier to set up, solve, and interpret complicated mathematical models of the real world.

Common Perspectives in Numerical Analysis

Numerical analysis is concerned with all aspects of the numerical solution of a problem, from the theoretical development and understanding of numerical methods to their practical implementation as reliable and efficient computer programs. Most numerical analysts specialize in small subfields, but they share some common concerns, perspectives, and mathematical methods of analysis. These include the following:

1. When presented with a problem that cannot be solved directly, they try to replace it with a "nearby problem" that can be solved more easily. Examples are the use of interpolation in developing numerical integration methods and root-finding methods.

2. There is widespread use of the language and results of linear algebra, real analysis, and functional analysis (with its simplifying notation of norms, vector spaces, and operators).

3. There is a fundamental concern with error, its size, and its analytic form. When approximating a problem, it is prudent to understand the nature of the error in the computed solution. Moreover, understanding the form of the error allows creation of extrapolation processes to improve the convergence behaviour of the numerical method.

4. Numerical analysts are concerned with stability, a concept referring to the sensitivity of the solution of a problem to small changes in the data or the parameters of the problem. Consider the following example. The polynomial,

$$p(x) = (x - 1)(x - 2)(x - 3)(x - 4)(x - 5)(x - 6)(x - 7),$$

or expanded,

$$p(x) = x^7 - 28x^6 + 322x^5 - 1,960x^4 - 6,769x^3 - 13,132x^2 + 13,068x - 5,040$$

has roots that are very sensitive to small changes in the coefficients. If the coefficient of x^6 is changed to -28.002, then the original roots 5 and 6 are perturbed to the complex numbers 5.459 0.540i—a very significant change in values. Such a polynomial $p(x)$ is called unstable or ill-conditioned with respect to the root-finding problem. Numerical methods for solving problems should be no more sensitive to changes in the data than the original problem to be solved. Moreover, the formulation of the original problem should be stable or well-conditioned.

5. Numerical analysts are very interested in the effects of using finite precision computer arithmetic. This is especially important in numerical linear algebra, as large problems contain many rounding errors.

6. Numerical analysts are generally interested in measuring the efficiency (or "cost") of an algorithm. For example, the use of Gaussian elimination to solve a linear system $Ax = b$ containing n equations will require approximately $2n^3/3$ arithmetic operations. Numerical analysts would want to know how this method compares with other methods for solving the problem.

Different Areas and Methods under Numerical Analysis

The field of numerical analysis is divided into different disciplines according to the problem that is to be solved.

One of the simplest problems is the evaluation of a function at a given point. The most straightforward approach, of just plugging in the number in the formula is sometimes not very efficient. For polynomials, a better approach is using the Horner scheme, since it reduces the necessary number of multiplications and additions. Generally, it is important to estimate and control round-off errors arising from the use of floating point arithmetic.

Interpolation, Extrapolation and Regression

Interpolation solves the following problem: given the value of some unknown function at a number of points, what value does that function have at some other point between the given points?

Extrapolation is very similar to interpolation, except that now we want to find the value of the unknown function at a point which is outside the given points.

Regression is also similar, but it takes into account that the data is imprecise. Given some points, and a measurement of the value of some function at these points (with an error), we want to determine the unknown function. The least squares-method is one popular way to achieve this.

Solving Equations and Systems of Equations

Another fundamental problem is computing the solution of some given equation. Two cases are commonly distinguished, depending on whether the equation is linear or not. For instance, the equation $2x + 5 = 3$ is linear while $2 \times 2 + 5 = 3$ is not.

Much effort has been put in the development of methods for solving systems of linear equations. Standard direct methods, i.e., methods that use some matrix decomposition are Gaussian elimination, LU decomposition, Cholesky decomposition for symmetric (or hermitian) and positive-definite matrix, and QR decomposition for non-square matrices. Iterative methods such as the Jacobi method, Gauss-Seidel method, successive over-relaxation and conjugate gradient method are usually preferred for large systems.

Root-finding algorithms are used to solve nonlinear equations (they are so named since a root of a function is an argument for which the function yields zero). If the function is differentiable and the derivative is known, then Newton's method is a popular choice. Linearization is another technique for solving nonlinear equations.

Solving Eigenvalue or Singular Value Problems

Several important problems can be phrased in terms of eigenvalue decompositions or singular value decompositions. For instance, the spectral image compression algorithm is based on the singular value decomposition. The corresponding tool in statistics is calledprincipal component analysis.

Optimization

Optimization problems ask for the point at which a given function is maximized (or minimized). Often, the point also has to satisfy some constraints.

The field of optimization is further split in several subfields, depending on the form of the objective function and the constraint. For instance, linear programming deals with the case that both the objective function and the constraints are linear. A famous method in linear programming is the simplex method.

The method of Lagrange multipliers can be used to reduce optimization problems with constraints to unconstrained optimization problems.

Evaluating Integrals

Numerical integration, in some instances also known as numerical quadrature, asks for the value of a definite integral. Popular methods use one of the Newton-Cotes formulas (like the midpoint rule or Simpson's rule) or Gaussian quadrature. These methods rely on a "divide and conquer" strategy, whereby an integral on a relatively large set is broken down into integrals on smaller sets. In higher dimensions, where these methods become prohibitively expensive in terms of computational effort, one may use Monte Carlo or quasi-Monte Carlo methods, or, in modestly large dimensions, the method of sparse grids.

Differential Equations

Numerical analysis is also concerned with computing (in an approximate way) the solution of differential equations, both ordinary differential equations and partial differential equations.

Partial differential equations are solved by first discretizing the equation, bringing it into a finite-dimensional subspace. This can be done by a finite element method, a finite differencemethod, or (particularly in engineering) a finite volume method. The theoretical justification of these methods often involves theorems from functional analysis.

Modern Applications and Computer Software

Numerical analysis and mathematical modelling are essential in many areas of modern life. Sophisticated numerical analysis software is commonly embedded in popular software packages (e.g., spreadsheet programs) and allows fairly detailed models to be evaluated, even when the user is unaware of the underlying mathematics. Attaining this level of user transparency requires reliable, efficient, and accurate numerical analysis software, and it requires problem-solving environments (PSE) in which it is relatively easy to model a given situation. PSEs are usually based on excellent theoretical mathematical models, made available to the user through a convenient graphical user interface.

Applications

Computer-aided engineering (CAE) is an important subject within engineering, and some quite sophisticated PSEs have been developed for this field. A wide variety of numerical analysis techniques is involved in solving such mathematical models. The models follow the basic Newtonian laws of mechanics, but there is a variety of possible specific models, and research continues on their design. One important CAE topic is that of modelling the dynamics of moving mechanical systems, a technique that involves both ordinary differential equations and algebraic equations (generally nonlinear). The numerical analysis of these mixed systems, called differential-algebraic systems, is quite difficult but necessary in order to model moving mechanical systems. Building simulators for cars, planes, and other vehicles requires solving differential-algebraic systems in real time.

Another important application is atmospheric modelling. In addition to improving weather forecasts, such models are crucial for understanding the possible effects of human activities on the Earth's climate. In order to create a useful model, many variables must be introduced. Fundamental among these are the velocity $V(x, y, z, t)$, pressure $P(x, y, z, t)$, and temperature $T(x, y, z, t)$, all given at position (x, y, z) and time t. In addition, various chemicals exist in the atmosphere, including ozone, certain chemical pollutants, carbon dioxide, and other gases and particulates, and their interactions have to be considered. The underlying equations for studying $V(x, y, z, t)$, $P(x, y, z, t)$, and $T(x, y, z, t)$ are partial differential equations; and the interactions of the various chemicals are described using some quite difficult ordinary differential equations. Many types of numerical analysis procedures are used in atmospheric modelling, including computational fluid mechanics and the numerical solution of differential equations. Researchers strive to include ever finer detail in atmospheric models, primarily by incorporating data over smaller and smaller local regions in the atmosphere and implementing their models on highly parallel supercomputers.

Modern businesses rely on optimization methods to decide how to allocate resources most efficiently. For example, optimization methods are used for inventory control, scheduling, determining the best location for manufacturing and storage facilities, and investment strategies.

Computer Software

Software to implement common numerical analysis procedures must be reliable, accurate, and efficient. Moreover, it must be written so as to be easily portable between different computer systems. Since about 1970, a number of government-sponsored research efforts have produced specialized, high-quality numerical analysis software.

The most popular programming language for implementing numerical analysis methods is Fortran, a language developed in the 1950s that continues to be updated to meet changing needs. Other languages, such as C, C++, and Java, are also used for numerical analysis. Another approach for basic problems involves creating higher level PSEs, which often contain quite sophisticated numerical analysis, programming, and graphical tools. Best known of these PSEs is MATLAB, a commercial package that is arguably the most popular way to do numerical computing. Two popular computer programs for handling algebraic-analytic mathematics (manipulating and displaying formulas) are Maple and Mathematica.

Numerical Linear Algebra

The study of algorithms aimed at the operation of linear algebra computations on computers is under the scope of numerical linear algebra. It has applications in image and signal processing, computational finance, telecommunication, materials science simulations, etc. This chapter has been carefully written to provide an easy understanding of numerical linear algebra through a detailed discussion of systems of linear equations, Gauss elimination method, pivoting, LU decomposition, Cholesky decomposition, etc.

A fundamentally important subject that deals with the theory and practice of processes in linear algebra. Principally these involve the central problems of the solution of linear algebraic equations $Ax = b$ and the eigenvalue problem in which eigenvalues λ_k and the eigenvectors x_k are sought where $Ax_k = \lambda_k x_k$.

Numerical linear algebra forms the basis of much scientific computing. Both of these problems have many variants, determined by the properties of the matrix A. For example, a related problem is the solution of over determined systems where A has more rows than columns. Here there are good reasons for computing x to minimize the norm $||Ax - b||_2$.

A major activity is the computing of certain linear transformations in the form of matrices, which brings about some simplification of the given problem. Most widely used are orthogonal matrices Q, for which $Q^T Q = I$.

An important feature of large-scale scientific computing is where the associated matrices are sparse, i.e. where a high proportion of the elements are zero. This is exploited in the algorithms for their solution.

The cross-fertilization between numerical linear algebra and real-life world has been very fruitful and continuously increasing in number and strength in the last decades. In particular, applications have been making increasingly sophisticated use of linear algebra on both theoretical and algorithmic fronts. The interaction between them has been growing, leading to many new algorithms.

On one hand, the need for "effective" (in the sense that can solve the problem but also in a relatively small wall clock time with the prescribed accuracy) algorithms is every day increasing, and the answers are every day more built around the given problem hybridizing various techniques. Our recent numerical methods for the large problems in the inner ear diseases. On the other hand, the issues related to sparse matrix software that are of interest to application scientists and industrial users are often fairly different from those on which the academic community is focused. For example, for an application scientist or an industrial user, improving robustness may be far more important than finding a method that would gain speed. Memory usage is also an important consideration, but is seldom accounted for in academic research on sparse matrix solvers.

As a last example, linear systems solved in applications are almost always part of some nonlinear iteration (e.g., Newton-like) or optimization loop. It is important to consider the coupling between the linear and nonlinear parts, instead of focusing on the linear systems alone.

Systems of Linear Equations

A linear equation in variables $x_1, x_2, ..., x_n$ is an equation of the form,

$$a_1 x_1 + a_2 x_2 + \cdots + a_n x_n = b,$$

where $a_1, a_2, ..., a_n$ and b are constant real or complex numbers. The constant a_i is called the coefficient of xi; and b is called the constant term of the equation.

A system of linear equations (or linear system) is a finite collection of linear equations in same variables. For instance, a linear system of m equations in n variables $x_1, x_2, ..., x_n$ can be written as,

$$\begin{cases} a_{11} x_1 + a_{12} x_2 + \cdots + a_{1n} x_n = b_1 \\ a_{21} x_1 + a_{22} x_2 + \cdots + a_{2n} x_n = b_2 \\ \quad\quad\quad\quad \vdots \\ a_{m1} x_1 + a_{m2} x_2 + \cdots + a_{mn} x_n = b_m \end{cases}$$

A solution of a linear system above is a tuple $(s_1, s_2, ..., s_n)$ of numbers that makes each equation a true statement when the values $s_1, s_2, ..., s_n$ are substituted for $x_1, x_2, ..., x_n$, respectively. The set of all solutions of a linear system is called the solution set of the system.

The augmented matrix of the general linear system (1:1) is the table,

$$\begin{bmatrix} a_{11} & a_{12} & \cdots & a_{1n} & b_1 \\ a_{21} & a_{22} & \cdots & a_{2n} & b_2 \\ \vdots & \vdots & \ddots & \vdots & \vdots \\ a_{m1} & a_{m2} & \cdots & a_{mn} & b_m \end{bmatrix}$$

and the coefficient matrix of (1:1) is,

$$\begin{bmatrix} a_{11} & a_{12} & \cdots & a_{1n} \\ a_{21} & a_{22} & \cdots & a_{2n} \\ \vdots & \vdots & \ddots & \vdots \\ a_{m1} & a_{m2} & \cdots & a_{mn} \end{bmatrix}$$

Elementary Row Operations

There are three kinds of elementary row operations on matrices:

(a) Adding a multiple of one row to another row;

(b) Multiplying all entries of one row by a nonzero constant;

(c) Interchanging two rows.

Row Echelon Forms

A matrix is said to be in row echelon form if it satisfies the following two conditions:

(a) All zero rows are gathered near the bottom.

(b) The first nonzero entry of a row, called the leading entry of that row, is ahead of the first nonzero entry of the next row.

   A matrix in row echelon form is said to be in reduced row echelon form if it satisfies two more conditions:

(c) The leading entry of every nonzero row is 1.

(d) Each leading entry 1 is the only nonzero entry in its column.

A matrix in (reduced) row echelon form is called a (reduced) row echelon matrix.

Sometimes we call row echelon forms just as echelon forms and row echelon matrices as echelon matrices without mentioning the word "row."

Row Echelon form Pattern

The following are two typical row echelon matrices.

$$\begin{bmatrix} \bullet & * & * & * & * & * & * & * & * \\ 0 & \bullet & * & * & * & * & * & * & * \\ 0 & 0 & 0 & 0 & \bullet & * & * & * & * \\ 0 & 0 & 0 & 0 & 0 & 0 & \bullet & * & * \\ 0 & 0 & 0 & 0 & 0 & 0 & 0 & 0 & 0 \\ 0 & 0 & 0 & 0 & 0 & 0 & 0 & 0 & 0 \end{bmatrix}, \quad \begin{bmatrix} 0 & \bullet & * & * & * & * & * & * & * \\ 0 & 0 & 0 & 0 & \bullet & * & * & * & * \\ 0 & 0 & 0 & 0 & 0 & 0 & \bullet & * & * \\ 0 & 0 & 0 & 0 & 0 & 0 & 0 & 0 & \bullet \\ 0 & 0 & 0 & 0 & 0 & 0 & 0 & 0 & 0 \\ 0 & 0 & 0 & 0 & 0 & 0 & 0 & 0 & 0 \end{bmatrix}$$

where the circled stars • represent arbitrary nonzero numbers, and the stars ✕ represent arbitrary numbers, including zero. The following are two typical reduced row echelon matrices.

$$\begin{bmatrix} 1 & 0 & * & * & 0 & * & 0 & * & * \\ 0 & 1 & * & * & 0 & * & 0 & * & * \\ 0 & 0 & 0 & 0 & 1 & * & 0 & * & * \\ 0 & 0 & 0 & 0 & 0 & 0 & 1 & * & * \\ 0 & 0 & 0 & 0 & 0 & 0 & 0 & 0 & 0 \\ 0 & 0 & 0 & 0 & 0 & 0 & 0 & 0 & 0 \end{bmatrix}, \begin{bmatrix} 0 & 1 & * & * & 0 & * & 0 & 0 & 0 \\ 0 & 0 & 0 & 0 & 1 & * & 0 & 0 & 0 \\ 0 & 0 & 0 & 0 & 0 & 0 & 1 & 0 & 0 \\ 0 & 0 & 0 & 0 & 0 & 0 & 0 & 0 & 1 \\ 0 & 0 & 0 & 0 & 0 & 0 & 0 & 0 & 0 \\ 0 & 0 & 0 & 0 & 0 & 0 & 0 & 0 & 0 \end{bmatrix}$$

If a matrix A is row equivalent to a row echelon matrix B, we say that A has the row echelon form B; if B is further a reduced row echelon matrix, then we say that A has the reduced row echelon form B.

Pivoting

The objective of pivoting is to make an element above or below a leading one into a zero. The "pivot" or "pivot element" is an element on the left hand side of a matrix that you want the elements above and below to be zero.

Pivot Process

Pivoting works because a common multiple (not necessarily the least common multiple) of two numbers can always be found by multiplying the two numbers together. Let's take the example we had before, and clear the first column.

$$\begin{array}{cccc} x & y & z & rhs \end{array}$$
$$\begin{bmatrix} 3 & 2 & -4 & 3 \\ 2 & 3 & 3 & 15 \\ 5 & -3 & 1 & 14 \end{bmatrix}$$

Selecting a Pivot

- Pick the column with the most zeros in it.

- Use a row or column only once

- Pivot on a one if possible

- Pivot on the main diagonal

- Never pivot on a zero

- Never pivot on the right hand side.

Since there is no one in the first row, we have two options: Either we divide the first row by three and work with fractions, or we pivot on the three and get large numbers. Go ahead and circle that as the pivot element. You may see the pivot elements with a * in front of it.

$$
\begin{array}{cccc}
x & y & z & \text{rhs} \\
\end{array}
$$
$$
\left[\begin{array}{ccc|c}
{}^*3 & 2 & -4 & 3 \\
2 & 3 & 3 & 15 \\
5 & -3 & 1 & 14
\end{array}\right]
$$

The idea is to make the boxed (yellow) numbers into zero. Using the combined row operation (this is not an elementary operation), that could be done by $3R_2 - 2R_1 \to R_2$ and $3R_3 - 5R_1 \to R_3$.

The only row not being changed is the row containing the pivot element (the 3). The whole point of the pivot process is to make the boxed values into zero. Go ahead and rewrite the pivot row and clear (make zero) the pivot column.

$$
\begin{array}{cccc}
x & y & z & \text{rhs} \\
\end{array}
$$
$$
\left[\begin{array}{ccc|c}
{}^*3 & 2 & -4 & 3 \\
0 & & & \\
0 & & &
\end{array}\right]
$$

To replace the values in row 2, each new element is obtained by multiplying the element being replaced in the second row by 3 and subtracting 2 times the element in the first row from the same column as the element being replaced.

To perform the pivot, place one finger on the pivot (circled number), and one finger on the element being replaced. Multiply these two numbers together. Now, place one finger on the boxed number in the same row as the element you're replacing and the other finger in the pivot row and the same column as the number your replacing. Multiply these two numbers together. Take the product with the pivot and subtract the product without the pivot.

$$
\begin{array}{cccc}
x & y & z & \text{rhs} \\
\end{array}
$$
$$
\left[\begin{array}{ccc|c}
{}^*3 & 2 & -4 & 3 \\
2 & 3 & 3 & 15 \\
5 & -3 & 1 & 14
\end{array}\right]
$$

To replace the 3 in R_2C_2, you would take 3(3) - 2(2) = 9 - 4 = 5.

To replace the 3 in R_2C_3, you would take 3(3) - 2(-4) = 9 +8 = 17.

To replace the 15 in R_2C_4, you would take 3(15) - 2(3) = 45 - 6 = 39.

To replace the -3 in R_3C_2, you would take 3(-3) - 5(2) = -9 - 10 = -19.

To replace the 1 in R_3C_3, you would take 3(1) - 5(-4) = 3 + 20 = 23

To replace the 14 in R_3C_4, you would take 3(14) - 5(3) = 42 - 15 = 27.

Here's how the process looks.

$$\begin{array}{cccc} x & y & z & rhs \end{array}$$

$$\left[\begin{array}{ccc|c} \text{pivot row, copy} & \text{pivot row, copy} & \text{pivot row, copy} & \text{pivot row, copy} \\ 3 & 2 & -4 & 3 \\ \text{pivot column, clear} & 3(3)-2(2) & 3(3)-2(-4) & 3(15)-2(3) \\ 0 & 5 & 17 & 39 \\ \text{pivot column, clear} & 3(-3)-5(2) & 3(1)-5(-4) & 3(14)-5(3) \\ 0 & -19 & 23 & 27 \end{array}\right]$$

Or, if you remove the comments, the matrix after the first pivot looks like this.

$$\begin{array}{cccc} x & y & z & rhs \end{array}$$
$$\left[\begin{array}{ccc|c} 3 & 2 & -4 & 3 \\ 0 & 5 & 17 & 39 \\ 0 & -19 & 23 & 27 \end{array}\right]$$

It is now time to repeat the entire process. We go through and pick another place to pivot. We would like it to be on the main diagonal, a one, or have zeros in the column. Unfortunately, we can't have any of those. But since we have to multiply all the other numbers by the pivot, we want it to be small, so we'll pivot on the 5 in R2C2 and clear out the 2 and -19.

$$\begin{array}{cccc} x & y & z & rhs \end{array}$$
$$\left[\begin{array}{ccc|c} 3 & 2 & -4 & 3 \\ 0 & *5 & 17 & 39 \\ 0 & -19 & 23 & 27 \end{array}\right]$$

Begin by copying down the pivot row (2nd row) and clearing the pivot column (2nd column). Previously cleared columns will remain cleared.

$$\begin{array}{cccc} x & y & z & rhs \end{array}$$

$$\begin{bmatrix} & 0 & & \\ 0 & {}^*5 & 17 & 39 \\ 0 & 0 & & \end{bmatrix}$$

Here are the calculations to find the next interation. Pay careful attention to the 3rd row where we're subtracting -19 times a value. Since we're subtracting a negative, We can write it as plus 19.

x	y	z	rhs
$5(3)-2(0)$	pivot column, clear	$5(-14)-2(17)$	$5(3)-2(39)$
15	0	-54	-63
pivot row, copy	pivot row, copy	pivot row, copy	pivot row, copy
0	5	17	39
previously cleared	pivot column, clear	$5(23)+19(17)$	$5(27)+19(39)$
0	0	438	876

And the resulting matrix.

$$\begin{array}{cccc} x & y & z & rhs \end{array}$$

$$\begin{bmatrix} 15 & 0 & -54 & -63 \\ 0 & 5 & 17 & 39 \\ 0 & 0 & 438 & 876 \end{bmatrix}$$

Notice that all the elements in the first row are multiples of 3 and all the elements in the last row are multiples of 438. We'll divide to reduce the rows.

$$\begin{array}{cccc} x & y & z & rhs \end{array}$$

$$\begin{bmatrix} 5 & 0 & -18 & -21 \\ 0 & 5 & 17 & 39 \\ 0 & 0 & 1 & 2 \end{bmatrix}$$

That had the added benefit of giving us a 1, exactly where we want it to be to pivot. So, we'll pivot on the 1 in R_3C_3 and clear out the -18 and 17. Circle your pivot and box the other numbers in that column to clear.

$$\begin{array}{cccc} x & y & z & rhs \end{array}$$

$$\begin{bmatrix} 5 & 0 & -18 & -21 \\ 0 & 5 & 17 & 39 \\ 0 & 0 & {}^*1 & 2 \end{bmatrix}$$

Copy down the pivot row and clear the pivot column. Previously cleared columns will remain cleared as long as you don't pivot in a row or column twice.

$$
\begin{array}{cccc}
x & y & z & rhs \\
\end{array}
$$

$$
\begin{bmatrix}
 & 0 & 0 & \\
0 & & 0 & \\
0 & 0 & *1 & 2
\end{bmatrix}
$$

Notice that each time, there are fewer calculations to perform. Here are the calculations for this pivot. Again, since the value in the pivot column in the first row is -18 and we're subtracting, we can write it as + 18.

$$
\begin{array}{cccc}
x & y & z & \\
\end{array}
$$

$$
\begin{bmatrix}
1(5)+18(0) & \text{previosly cleared} & \text{pivot column, clear} & 1(-21)+18(2) \\
5 & 0 & 0 & 15 \\
\text{previosly cleared} & 1(15)-17(0) & \text{pivot column, clear} & 1(39)-17(2) \\
0 & 5 & 0 & 5 \\
\text{pivot row, copy} & \text{pivot row, copy} & \text{pivot row, copy} & \text{pivot row, copy} \\
0 & 0 & 1 & 2
\end{bmatrix}
$$

And the resulting matrix.

$$
\begin{array}{cccc}
x & y & z & rhs \\
\end{array}
$$

$$
\begin{bmatrix}
1 & 0 & 0 & 3 \\
0 & 1 & 0 & 1 \\
0 & 0 & 1 & 2
\end{bmatrix}
$$

And the final answer is x = 3, y = 1, and z = 2. You can also write that as an ordered triplet {(3,1,2)}.

Pivoting Strategies

1. No pivoting. It means no row interchanges. It can be done only if Gaussian elimination never run into zeros on the diagonal. Since division by zero is a fatal error we usually avoid this pivoting strategy.

- Pivoting to Avoid $a_{p,p} = 0$

If $a_{p,p} = 0$, then row p cannot be used to eliminate the elements in column p below the main diagonal. It is necessary to find row k, where $a_{k,p} \neq 0$ and k > p, and then interchange row p and row k so that a nonzero pivot element is obtained. This process is called pivoting, and the criterion for deciding which row to choose is called a pivoting strategy. The first idea that comes to mind is the following one.

2. Trivial Pivoting. The trivial pivoting strategy is as follows. If $a_{p,p} \neq 0$, do not switch rows. If $a_{p,p} = 0$, locate the first row below p in which $a_{k,p} \neq 0$ and then switch rows k and p. This will result in a new element $a_{p,p} \neq 0$, which is a nonzero pivot element.

- Pivoting to Reduce Error

Because the computer uses fixed-precision arithmetic, it is possible that a small error will be introduced each time that an arithmetic operation is performed. The following example illustrates how use of the trivial pivoting strategy in Gaussian elimination can lead to significant error in the solution of a linear system of equations.

3. Partial Pivoting. The partial pivoting strategy is as follows. If $a_{p,p} \neq 0$, do not switch rows. If $a_{p,p} = 0$, locate row u below p in which $\left| a_{u,p} \right| = \max_{p+1 \leq i \leq n} \left| a_{i,p} \right|$ and $a_{u,p} \neq 0$ and then switch rows u and p. This will result in a new $a_{p,p} \neq 0$, which is a nonzero pivot element.

Remark: Only row permutations are permitted. The strategy is to switch the largest entry in the pivot column to the diagonal.

4. Scaled Partial Pivoting. At the start of the procedure we compute scale factors for each row of the matrix A as follows:

$$S_i = \max_{1 \leq i \leq n} \left| a_{i,j} \right| \text{ for } i=1, 2, \ldots, n .$$

The scale factors are interchanged with their corresponding row in the elimination steps.

The scaled partial pivoting strategy is as follows. If $a_{p,p} \neq 0$, do not switch rows. If $a_{p,p} = 0$, locate row u below p in which $\dfrac{\left| a_{u,p} \right|}{S_u} = \max_{p+1 \leq i \leq n} \dfrac{\left| a_{i,p} \right|}{S_i}$ and $a_{u,p} \neq 0$ and then switch rows u and p. This will result in a new element $a_{p,p} \neq 0$, which is a nonzero pivot element.

Remark: Only row permutations are permitted. The strategy is to switch the largest scaled entry in the pivot column to the diagonal.

5. Total Pivoting. The total pivoting strategy is as follows. If $a_{p,p} \neq 0$, do not switch rows. If $a_{p,p} = 0$, locate row u below p and column v to the right of p in which $\left| a_{u,v} \right| = \max_{p+1 \leq, j \leq n} \left| a_{i,j} \right|$ and $a_{u,v} \neq 0$ and then: first switch rows u and p and second switch column v and p. This will result in a new element $a_{p,p} \neq 0$, which is a nonzero pivot element. This is also called "complete pivoting" or "maximal pivoting."

Remark: Both row and column permutations are permitted. The strategy is to switch the largest entry in the part of the matrix that we have not yet processed to the diagonal.

Gauss Elimination Method

Gauss Jordan Elimination Method is a method to solve large linear equation numerically. It is done by manipulating the given matrix using elementary row operations. It puts zero both above and below each pivot element as it goes from top row of the matrix to the bottom. This method is also known as Back Substitution. This method is also known as Gaussian elimination method.

Gauss Jordan elimination is an algorithm for getting matrices in reduced row echelon form using elementary row operations. Gauss Jordan is well suited for solving linear systems.

Consider n linear equations to be solved,

$$a_{11}x_1 + a_{12}x_2 + \ldots + a_{1n}X_n = b_1$$
$$a_{21}x_1 + a_{22}x_2 + \ldots + a_{2n}X_n = b_2$$
$$\ldots\ldots\ldots\ldots\ldots\ldots\ldots\ldots\ldots$$
$$a_{n1}x_1 + a_{n2}x_2 + \ldots + a_{nn}X_n = b_n$$

If $\begin{bmatrix} a_{11} & a_{12} & . & . & a_{1n} & b_1 \\ a_{21} & a_{22} & . & . & a_{2n} & b_2 \\ . & . & & . & . & . \\ . & . & . & . & . & . \\ a_{n1} & a_{n2} & . & . & a_{nn} & b_n \end{bmatrix}$ is augmented matrix associated with a linear system, then

elementary row operations are used to reduce this matrix to $\begin{bmatrix} 1 & 0 & . & . & 0 & m_1 \\ 0 & 1 & . & . & . & m_2 \\ . & . & . & . & . & . \\ . & . & . & . & . & . \\ 0 & 0 & . & . & 1 & m_n \end{bmatrix}$

The solution appears in last column, i.e. $x_i = m_i$.

Gauss Jordan Elimination Step by Step

In order to solve a system of three linear equations by Gauss Jordan elimination method, elementary row operation has to be performed on the augmented matrix as follow:

Step 1:

- Transform the element at a_{11} position to 1, by a suitable elementary row transformation using the element at a_{21} or a_{31} position or otherwise.

- Transform the non-zero elements, if any, at a_{21}, a_{31} position as zeros (other elements of the first column) by using the element 1 at a_{11} position. If at the end of step 1, there is a non-zero element at a_{22} or a_{32} position, go to step 2. Otherwise skip it.

Step 2:

- Transform the element at a_{22} position as 1 by a suitable elementary row transformation using the element at a_{32} position or otherwise.

- Transform the other non-zero elements, if any, of the second column (i.e., the non-zero elements, if any, at a_{12} and a_{32} positions) as zeros, by using the element 1 at a_{22} position. At the end of step 2 or after skipping it for reasons specified above, examine the element at a_{33} position. If it is non-zero, go to step3.

Step 3:

- Transform the element at a_{33} position as 1 by dividing R_3 with a suitable number.

- Transform the other non-zero elements if any of the third column (i.e., the non-zero elements, if any, at a_{13}, a_{23} position) as zeros by using the 1 present at a_{33} position.

Example: Solve the following equations by Gauss Jordan Elimination method.

$$x + 2y = 3$$

$$-x - 2z = -5$$

$$-3x - 5y + z = -4$$

Solution:

Given system of linear equations is $x + 2y = 3$,

$$-x - 2z = -5$$

$$-3x - 5y + z = -4$$

Augmented matrix is,

$$\begin{bmatrix} 1 & 2 & 0 & | & 3 \\ -1 & 0 & -2 & | & -5 \\ -3 & -5 & 1 & | & -4 \end{bmatrix}$$

Using operations $R_2 \sim R_2 + R_1$ and $R_3 \sim R_3 + 3R_1$,

$$\begin{bmatrix} 1 & 2 & 0 & | & 3 \\ 0 & 2 & -2 & | & -2 \\ 0 & 1 & 1 & | & 5 \end{bmatrix}$$

$$R_1 \sim R_1 - R_2 \text{ and } R_2 = \frac{R_2}{2}$$

$$\begin{bmatrix} 1 & 0 & 0 & 5 \\ 0 & 1 & -1 & -1 \\ 0 & 1 & 1 & 5 \end{bmatrix}$$

$R_3 \sim R_3 - R_2$

$$\begin{bmatrix} 1 & 0 & 0 & 5 \\ 0 & 1 & -1 & -1 \\ 0 & 0 & 2 & 6 \end{bmatrix}$$

$R_3 \sim \dfrac{R_3}{2}$

$$\begin{bmatrix} 1 & 0 & 0 & 5 \\ 0 & 1 & -1 & -1 \\ 0 & 0 & 1 & 3 \end{bmatrix}$$

$R_1 \sim R_1 - 2R_3$ and $R_2 \sim R_2 + R_3$

Therefore, the solution of the system is x = -1, y = 2 and z = 3.

Example: Solve the following equations by Gauss Jordan Elimination method.

$3x + 4y + 5z = 18$

$2x - y + 8z = 13$

$5x - 2y + 7z = 20$

Solution:

The augmented Matrix is,

$$\begin{bmatrix} 3 & 4 & 5 & 18 \\ 2 & -1 & 8 & 13 \\ 5 & -2 & 7 & 20 \end{bmatrix}$$

On applying operation $R_1 \to R_1 - R_2$, we get,

$$\begin{bmatrix} 1 & 5 & -3 & 5 \\ 2 & -1 & 8 & 13 \\ 5 & -2 & 7 & 20 \end{bmatrix}$$

On applying $R_2 \to R_2 - 2R_1$, $R3 \to R_3 - 5R_1$, we get,

$$\begin{bmatrix} 1 & 5 & -3 & 5 \\ 0 & -11 & 14 & 3 \\ 0 & -27 & 22 & -5 \end{bmatrix}$$

On applying $R_2 \rightarrow -5R_2 + 2R_3$, we get,

$$\begin{bmatrix} 1 & 5 & -3 & 5 \\ 0 & 1 & -26 & -25 \\ 0 & -27 & 22 & -5 \end{bmatrix}$$

On applying $R_1 \rightarrow R_1 - 5R_2$, $R_3 \rightarrow R_3 + 27R_2$, we obtain,

$$\begin{bmatrix} 1 & 0 & 127 & 130 \\ 0 & 1 & -26 & -25 \\ 0 & 0 & -680 & -680 \end{bmatrix}$$

On applying $R_3 \rightarrow R_3 / (-680)$, we get,

$$\begin{bmatrix} 1 & 0 & 127 & 130 \\ 0 & 1 & -26 & -25 \\ 0 & 0 & 1 & 1 \end{bmatrix}$$

On applying $R_1 \rightarrow R_1 - 127 R_3$, $R_2 \rightarrow R_2 + 26 R_3$, we obtain,

$$\begin{bmatrix} 1 & 0 & 0 & 3 \\ 0 & 1 & 0 & 1 \\ 0 & 0 & 1 & 1 \end{bmatrix}$$

Hence, the solution is x = 3, y = 1, z = 1.

Example: Use Gaussian elimination to solve the system of linear equations,

$$2x_2 + x_3 = -8$$
$$x_1 - 2x_2 - 3_{x3} = 0$$
$$-x_1 + x_2 + 2x_3 = 3.$$

Solution: As before, we carry out reduction on the system of equations and on the augmented matrix simultaneously, in order to make it clear that row operations on equations correspond exactly to row operations on matrices.

$$2x_2 + x_3 = -8 \qquad\qquad \begin{pmatrix} 0 & 2 & 1 & -8 \\ 1 & -2 & -3 & 0 \\ -1 & 1 & 2 & 3 \end{pmatrix}$$
$$x_1 - 2x_2 - 3_{x3} = 0$$
$$-x_1 + x_2 + 2x_3 = 3.$$

Swap Row 1 and Row 2.

$$\begin{aligned} x_1 - 2x_2 - 3x_3 &= 0 \\ 2x_2 + x_3 &= -8 \\ -x_1 + x_2 + 2x_3 &= 3 \end{aligned} \qquad \left(\begin{array}{ccc|c} 1 & -2 & -3 & 0 \\ 0 & 2 & 1 & -8 \\ -1 & 1 & 2 & 3 \end{array}\right)$$

Add Row 1 to Row 3.

$$\begin{aligned} x_1 - 2x_2 - 3x_3 &= 0 \\ 2x_2 + x_3 &= -8 \\ -x_2 - x_3 &= 3 \end{aligned} \qquad \left(\begin{array}{ccc|c} 1 & -2 & -3 & 0 \\ 0 & 2 & 1 & -8 \\ 0 & -1 & -1 & 3 \end{array}\right)$$

Swap Row 2 and Row 3.

$$\begin{aligned} x_1 - 2x_2 - 3x_3 &= 0 \\ -x_2 - x_3 &= 3 \\ 2x_2 + x_3 &= -8 \end{aligned} \qquad \left(\begin{array}{ccc|c} 1 & -2 & -3 & 0 \\ 0 & -1 & -1 & 3 \\ 0 & 2 & 1 & -8 \end{array}\right)$$

Add twice Row 2 to Row 3.

$$\begin{aligned} x_1 - 2x_2 - 3x_3 &= 0 \\ -x_2 - x_3 &= 3 \\ -x_3 &= -2 \end{aligned} \qquad \left(\begin{array}{ccc|c} 1 & -2 & -3 & 0 \\ 0 & -1 & -1 & 3 \\ 0 & 2 & -1 & -2 \end{array}\right)$$

Add −1 times Row 3 to Row 2.

Add −3 times Row 3 to Row 1.

$$\begin{aligned} x_1 - 2x_2 &= 6 \\ -x_2 &= 5 \\ -x_3 &= -2 \end{aligned} \qquad \left(\begin{array}{ccc|c} 1 & -2 & 0 & 6 \\ 0 & -1 & 0 & 5 \\ 0 & 0 & -1 & -2 \end{array}\right)$$

Add −2 times Row 2 to Row 1.

$$\begin{aligned} x_1 &= -4 \\ -x_2 &= 5 \\ -x_3 &= -2 \end{aligned} \qquad \left(\begin{array}{ccc|c} 1 & 0 & 0 & -4 \\ 0 & -1 & 0 & 5 \\ 0 & 0 & -1 & -2 \end{array}\right)$$

Multiply Rows 2 and 3 by −1.

$$\begin{aligned} x_1 &= -4 \\ x_2 &= -5 \\ x_3 &= 2 \end{aligned} \qquad \left(\begin{array}{ccc|c} 1 & 0 & 0 & -4 \\ 0 & 1 & 0 & -5 \\ 0 & 0 & 1 & 2 \end{array}\right)$$

Computational Efficiency

The number of arithmetic operations required to perform row reduction is one way of measuring the algorithm's computational efficiency. For example, to solve a system of n equations for n unknowns by performing row operations on the matrix until it is in echelon form, and then solving for each unknown in reverse order, requires $\dfrac{n(n+1)}{2}$ divisions, $\dfrac{2n^3 + 3n^2 - 5n}{6}$ multiplications, and $\dfrac{2n^3 + 3n^2 - 5n}{6}$ subtractions, for a total of approximately $\dfrac{2n^3}{3}$ operations. Thus it has arithmetic complexity of $O(n^3)$. This arithmetic complexity is a good measure of the time needed for the whole computation when the time for each arithmetic operation is approximately constant. This is the case when the coefficients are represented by floating-point numbers or when they belong to a finite field. If the coefficients are integers or rational numbers exactly represented, the intermediate entries can grow exponentially large, so the bit complexity is exponential. However, there is a variant of Gaussian elimination, called Bareiss algorithm that avoids this exponential growth of the intermediate entries, and, with the same arithmetic complexity of $O(n^3)$, has a bit complexity of $O(n^5)$.

This algorithm can be used on a computer for systems with thousands of equations and unknowns. However, the cost becomes prohibitive for systems with millions of equations. These large systems are generally solved using iterative methods. Specific methods exist for systems whose coefficients follow a regular pattern.

To put an $n \times n$ matrix into reduced echelon form by row operations, one needs n^3 arithmetic operations, which is approximately 50% more computation steps.

One possible problem is numerical instability, caused by the possibility of dividing by very small numbers. If, for example, the leading coefficient of one of the rows is very close to zero, then to row-reduce the matrix one would need to divide by that number so the leading coefficient is 1. This means that any error existed for the number which was close to zero would be amplified. Gaussian elimination is numerically stable for diagonally dominant or positive-definite matrices. For general matrices, Gaussian elimination is usually considered to be stable, when using partial pivoting, even though there are examples of stable matrices for which it is unstable.

Generalizations

The Gaussian elimination can be performed over any field, not just the real numbers.

Gaussian elimination does not generalize in any way to higher-order tensors (matrices are array representations of order-2 tensors); even computing the rank of a tensor of order greater than 2 is NP-hard.

Pseudo Code

Gaussian elimination transforms a given $m \times n$ matrix A into a matrix in row-echelon form.

In the following pseudocode, A[i, j] denotes the entry of the matrix A in row i and column j with the indices starting from 1. The transformation is performed *in place*, meaning that the original matrix is lost for being eventually replaced by its row-echelon form.

```
h := 1 /* Initialization of the pivot row */

k := 1 /* Initialization of the pivot column */

 while h ≤ m and k ≤ n

  /* Find the k-th pivot: */

  i_max := argmax (i = h ... m, abs(A[i, k]))

  if A[i_max, k] = 0

   /* No pivot in this column, pass to next column */

   k := k+1

 else

   swap rows(h, i_max)

   /* Do for all rows below pivot: */

   for i = h + 1 ... m:

    f := A[i, k] / A[h, k]

    /* Fill with zeros the lower part of pivot column: */

    A[i, k] := 0

    /* Do for all remaining elements in current row: */

    for j = k + 1 ... n:

     A[i, j] := A[i, j] - A[h, j] * f

   /* Increase pivot row and column */

   h := h+1

   k := k+1
```

This algorithm differs slightly from the one discussed earlier, by choosing a pivot with largest absolute value. Such a *partial pivoting* may be required if, at the pivot place, the entry of the matrix is zero. In any case, choosing the largest possible absolute value of the pivot improves the numerical stability of the algorithm, when floating point is used for representing numbers.

Upon completion of this procedure the matrix will be in row echelon form and the corresponding system may be solved by back substitution.

Applications

The common applications of Gauss Jordan elimination method are discussed below.

Solving System of Linear Equations

The first and most important application of Gauss-Jordan elimination method is for finding the solution of systems of linear equations which is applied throughout the mathematics.

Finding Determinant

The Gaussian elimination can be applied to a square matrix in order to find determinant of the matrix.

Finding Inverse of Matrix

The Gauss-Jordan elimination method can be used in determining the inverse of a square matrix. If M be an n x n square matrix and its inverse exists (since it is not necessary that each matrix has inverse matrix), then row reduction method can be used to compute inverse of such matrix.

Finding Ranks and Bases

The ranks as well as bases of square matrices can be computed by Gaussian elimination method using reduced row echelon form.

LU Decomposition

L U decomposition of a matrix is the factorization of a given square matrix into two triangular matrices, one upper triangular matrix and one lower triangular matrix, such that the product of these two matrices gives the original matrix. It was introduced by Alan Turing in 1948, who also created the Turing machine.

This method of factorizing a matrix as a product of two triangular matrices has various applications such as solution of a system of equations, which itself is an integral part of many applications such as finding current in a circuit and solution of discrete dynamical system problems; finding the inverse of a matrix and finding the determinant of the matrix.

Basically, the L U decomposition method comes handy whenever it is possible to model the problem to be solved into matrix form. Conversion to the matrix form and solving with triangular matrices makes it easy to do calculations in the process of finding the solution.

A square matrix A can be decomposed into two square matrices L and U such that A = L U where U is an upper triangular matrix formed as a result of applying Gauss Elimination Method on A; and L is a lower triangular matrix with diagonal elements being equal to 1.

For,

$$A = \begin{bmatrix} a_{11} & a_{12} & a_{13} \\ a_{21} & a_{22} & a_{23} \\ a_{31} & a_{32} & a_{33} \end{bmatrix}, \text{ we have } L = \begin{bmatrix} 1 & 0 & 0 \\ l_{21} & 1 & 0 \\ l_{31} & l_{32} & 1 \end{bmatrix} \text{ and } U = \begin{bmatrix} u_{11} & u_{12} & u_{13} \\ 0 & u_{22} & u_{23} \\ 0 & 0 & u_{33} \end{bmatrix} ; \text{ such that } A = L U.$$

Method of LU Decomposition:

Step 1: The system AX = B can be written as LUX = B.

Step 2: Define the matrix: n × 1 matrix y by UX = Y.

Step 3: The system LUX = B can be written as LY = B by using UX = Y.

Step 4: Solve the system of equations for y.

Step 5: Substitute y in UX = y.

Step 6: Solve for x.

General representation of LU decomposition of a 4×4 matrix:

$$\begin{bmatrix} l_{11} & 0 & 0 & 0 \\ l_{21} & l_{22} & 0 & 0 \\ l_{31} & l_{32} & l_{33} & 0 \\ l_{41} & l_{42} & l_{43} & l_{44} \end{bmatrix} \begin{bmatrix} u_{11} & u_{12} & u_{13} & u_{14} \\ 0 & u_{22} & u_{23} & u_{24} \\ 0 & 0 & u_{33} & u_{34} \\ 0 & 0 & 0 & u_{44} \end{bmatrix} = \begin{bmatrix} a_{11} & a_{12} & a_{13} & a_{14} \\ a_{21} & a_{22} & a_{23} & a_{24} \\ a_{31} & a_{32} & a_{33} & a_{34} \\ a_{41} & a_{42} & a_{43} & a_{44} \end{bmatrix}$$

Example:

Consider the matrix,

$$A = \begin{bmatrix} 2 & 1 \\ -1 & 3 \end{bmatrix}$$

The matrix A is of the form A=LU. That is,

$$A = LU$$

$$= \begin{bmatrix} 1 & 0 \\ l_{21} & 1 \end{bmatrix} \begin{bmatrix} u_{11} & u_{12} \\ 0 & u_{22} \end{bmatrix}$$

$$= \begin{bmatrix} u_{11} & u_{12} \\ l_{21}u_{11} & l_{21}u_{12} + u_{22} \end{bmatrix}$$

Therefore, $\begin{bmatrix} 2 & 1 \\ -1 & 3 \end{bmatrix} = \begin{bmatrix} u_{11} & u_{12} \\ l_{21}u_{11} & l_{21}u_{12} + u_{22} \end{bmatrix}$

Compare the entries of the matrix on both the sides.

$$\begin{cases} u_{11} = 2 \\ u_{12} = 1 \\ l_{21}u_{11} = -1 \\ l_{21}u_{12} + u_{22} = 3 \end{cases}$$

Solve the equations and obtain the results as,

$$u_{11} = 2, u_{12} = 1, l_{21} = -\frac{1}{2} \text{ and } u_{22} = \frac{7}{2}$$

Thus, the decomposition of the matrix A is,

$$\begin{bmatrix} 2 & 1 \\ -1 & 3 \end{bmatrix} = \begin{bmatrix} 1 & 0 \\ -\dfrac{1}{2} & 1 \end{bmatrix} \begin{bmatrix} 2 & 1 \\ 0 & \dfrac{7}{2} \end{bmatrix}$$

Example: Let

$$A = \begin{bmatrix} 2 & -3 & 1 \\ 1 & 2 & -3 \\ 4 & -1 & -2 \end{bmatrix}.$$

Express A as A = LU, where L and U are lower and upper triangular matrices and hence solve the system of equations:

$$2x_1 - 3x_2 + x_3 = 1, \ x_1 + 2x_2 - 3x_3 = 4, \ 4x_1 - x_2 - 2x_3 = 8.$$

Also, determine L^{-1}, U^{-1}, A^{-1} and $|A|$.

Solution: Let

$$\begin{bmatrix} 2 & -3 & 1 \\ 1 & 2 & -3 \\ 4 & -1 & -2 \end{bmatrix}$$

$$= \begin{bmatrix} l_{11} & 0 & 0 \\ l_{21} & l_{22} & 0 \\ l_{31} & l_{32} & l_{33} \end{bmatrix} \begin{bmatrix} 1 & u_{12} & u_{13} \\ 0 & 1 & u_{23} \\ 0 & 0 & 1 \end{bmatrix} = \begin{bmatrix} l_{11} & l_{11}u_{12} & l_{11}u_{13} \\ l_{21} & l_{21}u_{12}+l_{22} & l_{21}u_{13}+l_{22}u_{23} \\ l_{31} & l_{31}u_{12}+l_{32} & l_{31}u_{13}+l_{32}u_{23}+l_{33} \end{bmatrix}.$$

To find the values of l_{ij} and u_{ij}, comparing both sides and we obtained,

$$l_{11} = 2, \; l_{21} = 1 \qquad l_{31} = 4$$
$$l_{11}u_{12} = -3 \qquad \text{or,} \quad u_{12} = -3/2$$
$$l_{11}u_{13} = 1 \qquad \text{or,} \quad u_{13} = 1/2$$
$$l_{21}u_{12}+l_{22} = 2 \qquad \text{or,} \quad l_{22} = 7/2$$
$$l_{31}u_{12}+l_{32} = -1 \qquad \text{or,} \quad l_{32} = 7$$
$$l_{21}u_{13}+l_{22}u_{23} = -3 \qquad \text{or,} \quad u_{23} = -1$$
$$l_{31}u_{13}+l_{32}u_{23}+l_{33} = -2 \qquad \text{or,} \quad l_{33} = 1.$$

Hence L and U are given by,

$$L = \begin{bmatrix} 2 & 0 & 0 \\ 1 & 7/2 & 0 \\ 4 & 5 & 1 \end{bmatrix}, \; U = \begin{bmatrix} 1 & -3/2 & 1/2 \\ 0 & 1 & -1 \\ 0 & 0 & 1 \end{bmatrix}$$

The given equations can be written as Ax = b, where

$$A = \begin{bmatrix} 2 & -3 & 1 \\ 1 & 2 & -3 \\ 4 & -1 & -2 \end{bmatrix}, \; x = \begin{bmatrix} x_1 \\ x_2 \\ x_3 \end{bmatrix}, \; b = \begin{bmatrix} 1 \\ 4 \\ 8 \end{bmatrix}.$$

Let A = LU. Then, LUx = b. Let Ux = z. Then, the given equation reduces to Lz = b.

First we consider the equation Lz = b. Then

$$\begin{bmatrix} 2 & 0 & 0 \\ 1 & 7/2 & 0 \\ 4 & 5 & 1 \end{bmatrix} \begin{bmatrix} z_1 \\ z_2 \\ z_3 \end{bmatrix} = \begin{bmatrix} 1 \\ 4 \\ 8 \end{bmatrix}.$$

In explicit form these equations are,

$$2z_1 \qquad\qquad = 1,$$
$$z_1 + 7/2z_2 \qquad = 4,$$
$$4z_1 + 5z_2 + z_3 = 8.$$

The solution of the above equations is $z_1 = 1/2$; $z_2 = 1$; $z_3 = 1$.

Therefore, $z = (1/2; 1; 1)^t$.

Now, we solve the equation $Ux = z$, i.e.

$$\begin{bmatrix} 1 & -3/2 & 1/2 \\ 0 & 1 & -1 \\ 0 & 0 & 1 \end{bmatrix}\begin{bmatrix} x_1 \\ x_2 \\ x_3 \end{bmatrix} = \begin{bmatrix} 1/2 \\ 1 \\ 1 \end{bmatrix}$$

In explicit form, the equations are,

$$x_1 - (3/2)x_2 + (1/2)x_3 = 1/2$$
$$x_2 - x_3 = 1$$
$$x_3 = 1.$$

The solution is $x_3 = 1$; $x_2 = 1 + 1 = 2$; $x_1 = 1/2 + (3/2) x_2 - (1/2) x_3 = 3$, i.e.

$x_1 = 3$; $x_2 = 2$; $x_3 = 1$.

Third Part. Gauss-Jordan method is used to find L^{-1}. Augmented matrix is

$$[L\vdots I] = \begin{bmatrix} 2 & 0 & 0 & \vdots & 1 & 0 & 0 \\ 1 & 7/2 & 0 & \vdots & 0 & 1 & 0 \\ 4 & 5 & 1 & \vdots & 0 & 0 & 1 \end{bmatrix}$$

$$\sim \begin{bmatrix} 1 & 0 & 0 & \vdots & 1/2 & 0 & 0 \\ 1 & 7/2 & 0 & \vdots & 0 & 1 & 0 \\ 4 & 5 & 1 & \vdots & 0 & 0 & 1 \end{bmatrix} R_1' = \frac{1}{2}R_1$$

$$\sim \begin{bmatrix} 1 & 0 & 0 & \vdots & 1/2 & 0 & 0 \\ 0 & 7/2 & 0 & \vdots & -1/2 & 1 & 0 \\ 0 & 5 & 1 & \vdots & -2 & 0 & 1 \end{bmatrix} R_2' = R_2 - R_1, R_3' = R_3 - 4R_1$$

$$\sim \begin{bmatrix} 1 & 0 & 0 & \vdots & 1/2 & 0 & 0 \\ 0 & 1 & 0 & \vdots & -1/7 & 2/70 \\ 0 & 5 & 1 & \vdots & -2 & 0 & 1 \end{bmatrix} = R_1' = \frac{2}{7}R_2$$

$$\sim \begin{bmatrix} 1 & 0 & 0 & \vdots & 1/2 & 0 & 0 \\ 0 & 1 & 0 & \vdots & -1/7 & 2/7 & 0 \\ 0 & 0 & 1 & \vdots & -9/7 & -10/7 & 1 \end{bmatrix} R_3' = R_3 - 5R_2.$$

Thus, $L^{-1} = \begin{bmatrix} 1/2 & 0 & 0 \\ -1/7 & 2/7 & 0 \\ -9/7 & -10/7 & 1 \end{bmatrix}$

Using same process, one can determine U^{-1}. But, here another method is used to determine U^{-1}. We know that the inverse of an upper triangular matrix is upper triangular.

Therefore, let $U^{-1} = \begin{bmatrix} 1 & b_{12} & b_{13} \\ 0 & 1 & b_{23} \\ 0 & 0 & 1 \end{bmatrix}$

From the identity $U^{-1}U = I$, we have

$$\begin{bmatrix} 1 & b_{12} & b_{13} \\ 0 & 1 & b_{23} \\ 0 & 0 & 1 \end{bmatrix}\begin{bmatrix} 1 & -3/2 & 1/2 \\ 0 & 1 & -1 \\ 0 & 0 & 1 \end{bmatrix} = \begin{bmatrix} 1 & 0 & 0 \\ 0 & 1 & 0 \\ 0 & 0 & 1 \end{bmatrix}.$$

This gives, $\begin{bmatrix} 1 & -3/2+b_{12} & 1/2-b_{12}+b_{13} \\ 0 & 1 & -1+b_{23} \\ 0 & 0 & 1 \end{bmatrix} = \begin{bmatrix} 1 & 0 & 0 \\ 0 & 1 & 0 \\ 0 & 0 & 1 \end{bmatrix}.$

Comparing both sides,

-3=2 + b$_{12}$ = 0 or, b$_{12}$ = 3/2, 1/2 -b$_{12}$ + b$_{13}$ = 0 or, b$_{13}$ = 1

-1 + b$_{23}$ = 0 or, b$_{23}$ = 1

Thus,

$$U^{-1} = \begin{bmatrix} 1 & 3/2 & 1 \\ 0 & 1 & 1 \\ 0 & 0 & 1 \end{bmatrix}.$$

Now,

$$A^{-1} = U^{-1}L^{-1} = \begin{bmatrix} 1 & 3/2 & 1 \\ 0 & 1 & 1 \\ 0 & 0 & 1 \end{bmatrix}\begin{bmatrix} 1/2 & 0 & 0 \\ -1/7 & 2/7 & 0 \\ -9/7 & -10/7 & 1 \end{bmatrix}$$

$$= \begin{bmatrix} -1 & -1 & 1 \\ -10/7 & -8/7 & 1 \\ -9/7 & -10/7 & 1 \end{bmatrix}.$$

Last Part. $|A| = |L||U| = (2\times(7/2)\times1)\times1 = 7.$

LU Factorization with Partial Pivoting

It turns out that a proper permutation in rows (or columns) is sufficient for the LU factorization. The LU factorization with partial pivoting (LUP) refers often to the LU factorization with row permutations only:

$$PA = LU,$$

where L and U are again lower and upper triangular matrices, and P is a permutation matrix, which, when left-multiplied to A, reorders the rows of A. It turns out that all square matrices can be factorized in this form, and the factorization is numerically stable in practice. This makes LUP decomposition a useful technique in practice.

LU Factorization with Full Pivoting

An LU factorization with full pivoting involves both row and column permutations:

$$PAQ = LU,$$

where L, U and P are defined as before, and Q is a permutation matrix that reorders the columns of A.

LDU Decomposition

An LDU decomposition is a decomposition of the form,

$$A = LDU,$$

where D is a diagonal matrix, and L and U are *unit* triangular matrices, meaning that all the entries on the diagonals of L and U are one.

Above we required that A be a square matrix, but these decompositions can all be generalized to rectangular matrices as well. In that case, L and D are square matrices both of which have the same number of rows as A, and U has exactly the same dimensions as A. *Upper triangular* should be interpreted as having only zero entries below the main diagonal, which starts at the upper left corner.

Example:

We factorize the following 2-by-2 matrix:

$$\begin{bmatrix} 4 & 3 \\ 6 & 3 \end{bmatrix} = \begin{bmatrix} l_{11} & 0 \\ l_{21} & l_{22} \end{bmatrix} \begin{bmatrix} u_{11} & u_{12} \\ 0 & u_{22} \end{bmatrix}.$$

One way to find the LU decomposition of this simple matrix would be to simply solve the linear equations by inspection. Expanding the matrix multiplication gives,

$$l_{11} \cdot u_{11} + 0 \cdot 0 = 4$$
$$l_{11} \cdot u_{12} + 0 \cdot u_{22} = 3$$
$$l_{21} \cdot u_{11} + l_{22} \cdot 0 = 6$$
$$l_{21} \cdot u_{12} + l_{22} \cdot u_{22} = 3.$$

This system of equations is underdetermined. In this case any two non-zero elements of L and U matrices are parameters of the solution and can be set arbitrarily to any non-zero value. Therefore, to find the unique LU decomposition, it is necessary to put some restriction on L and U matrices. For example, we can conveniently require the lower triangular matrix L to be a unit triangular matrix (i.e. set all the entries of its main diagonal to ones). Then the system of equations has the following solution:

$$l_{21} = 1.5$$
$$u_{11} = 4$$
$$u_{12} = 3$$
$$u_{22} = -1.5$$

Substituting these values into the LU decomposition above yields,

$$\begin{bmatrix} 4 & 3 \\ 6 & 3 \end{bmatrix} = \begin{bmatrix} 1 & 0 \\ 1.5 & 1 \end{bmatrix} \begin{bmatrix} 4 & 3 \\ 0 & -1.5 \end{bmatrix}.$$

Existence and Uniqueness

Square Matrices

Any square matrix A admits an LUP factorization. If A is invertible, then it admits an LU (or LDU) factorization if and only if all its leading principal minors are nonzero. If A is a singular matrix of rank k, then it admits an LU factorization if the first k leading principal minors are nonzero, although the converse is not true.

If a square, invertible matrix has an LDU factorization with all diagonal entries of L and U equal to 1, then the factorization is unique. In that case, the LU factorization is also unique if we require that the diagonal of L (or U) consists of ones.

Symmetric Positive Definite Matrices

If A is a symmetric (or Hermitian, if A is complex) positive definite matrix, we can arrange matters so that U is the conjugate transpose of L. That is, we can write A as,

$$A = LL^*$$

This decomposition is called the Cholesky decomposition. The Cholesky decomposition always exists and is unique — provided the matrix is positive definite. Furthermore, computing the Cholesky decomposition is more efficient and numerically more stable than computing some other LU decompositions.

General Matrices

For a (not necessarily invertible) matrix over any field, the exact necessary and sufficient conditions under which it has an LU factorization are known. The conditions are expressed in terms of the ranks of certain submatrices. The Gaussian elimination algorithm for obtaining LU decomposition has also been extended to this most general case.

Example: Solve the following system of linear equations, by Crout's method:

$$10x_1 + 3x_2 + 4x_3 = +15$$
$$2x_1 - 10x_2 + 3x_3 = 37$$
$$3x_1 + 2x_2 - 10x_3 = -10$$

Solution: In matrix form, the given system of equation can be written as,

$$\begin{pmatrix} 10 & 3 & 4 \\ 2 & -10 & 3 \\ 3 & 2 & -10 \end{pmatrix} \begin{pmatrix} x_1 \\ x_2 \\ x_3 \end{pmatrix} = \begin{pmatrix} 15 \\ 37 \\ -10 \end{pmatrix}$$

which is of the form $A\underset{\sim}{x} = \underset{\sim}{b}$. Let A = LU, which implies

$$\begin{pmatrix} 10 & 3 & 4 \\ 2 & -10 & 3 \\ 3 & 2 & -10 \end{pmatrix} = \begin{pmatrix} l_{11} & 0 & 0 \\ l_{21} & l_{22} & 0 \\ l_{31} & l_{32} & l_{33} \end{pmatrix} \begin{pmatrix} 1 & u_{12} & u_{13} \\ 0 & 1 & u_{23} \\ 0 & 0 & 1 \end{pmatrix}$$

$$= \begin{pmatrix} l_{11} & l_{11}u_{12} & l_{11}u_{13} \\ l_{21} & l_{21}u_{12}+l_{22} & l_{21}u_{13}+l_{22}u_{23} \\ l_{31} & l_{31}u_{12}+l_{32} & l_{31}u_{13}+l_{32}u_{23}+l_{33} \end{pmatrix}$$

$$\Rightarrow l_{11} = 10, l_{21} = 2, l_{31} = 3; u_{12}\frac{3}{10}, u_{13} = \frac{4}{10};$$

$$l_{21}u_{12}+l_{22} = -10 \Rightarrow l_{22} = -10-2\times\frac{3}{10} = -\frac{106}{10};$$

$$l_{21}u_{13}+l_{22}u_{23} = 3 \Rightarrow u_{23} = \frac{\left(3-2\times\dfrac{4}{10}\right)}{\left(-\dfrac{106}{10}\right)} = -\frac{11}{35};$$

$$l_{31}u_{12} + l_{32} = 2 \Rightarrow l_{32} = 2 - l_{31}u_{12} = 2 - 3 \times \frac{3}{10} = \frac{11}{10};$$

$$l_{31}u_{13} + l_{32}u_{23} + l_{33} = -10 \Rightarrow l_{33} = -10 - l_{31}u_{13} - l_{32}u_{23}$$

$$= -10 - 3 \times \frac{4}{10} + \frac{11}{10} \times \frac{11}{53} = -\frac{1163}{106}$$

Therefore, we get,

$$L = \begin{pmatrix} 10 & 0 & 0 \\ 2 & \dfrac{-106}{10} & 0 \\ 3 & \dfrac{11}{10} & \dfrac{-1163}{106} \end{pmatrix} \text{ and } U = \begin{pmatrix} 1 & \dfrac{3}{10} & \dfrac{4}{10} \\ 0 & 1 & \dfrac{-11}{53} \\ 0 & 0 & 1 \end{pmatrix}$$

Now, let $Ux = \underset{\sim}{y}$, then $L\underset{\sim}{y} = \underset{\sim}{b}$ implies,

$$\begin{pmatrix} 10 & 0 & 0 \\ 2 & \dfrac{-106}{10} & 0 \\ 3 & \dfrac{11}{10} & \dfrac{-1163}{106} \end{pmatrix} \begin{pmatrix} y_1 \\ y_2 \\ y_3 \end{pmatrix} = \begin{pmatrix} 15 \\ 37 \\ -10 \end{pmatrix}$$

$$10y_1 = 15 \Rightarrow y_1 = \frac{3}{2}$$

$$2y_1 - \frac{106}{10}y_2 = 37 \Rightarrow y_2 = \frac{-170}{53}$$

Thus, $y = \begin{pmatrix} \dfrac{3}{2} \\ \dfrac{-170}{53} \\ 1 \end{pmatrix}$ and $U\tilde{x} = \underset{\sim}{y}$ gives

$$\begin{pmatrix} 1 & \dfrac{3}{10} & \dfrac{4}{10} \\ 0 & 1 & \dfrac{-11}{53} \\ 0 & 0 & 1 \end{pmatrix} \begin{pmatrix} x_1 \\ x_2 \\ x_3 \end{pmatrix} = \begin{pmatrix} y_1 \\ y_2 \\ y_3 \end{pmatrix} = \begin{pmatrix} \dfrac{3}{2} \\ \dfrac{-170}{53} \\ 1 \end{pmatrix}, \text{ which implies}$$

$$x_1 + \frac{3}{10}x_2 - \frac{4}{10}x_3 = \frac{3}{2}$$

$$x_2 - \frac{11}{53}x_3 = \frac{-170}{53}$$

$$x_3 = 1$$

By back substitution, we get,

$$x_3 = 1$$

$$x_2 = \frac{11 \times 1}{53} - \frac{170}{53} = -3$$

$$x_1 = \frac{3}{2} - \frac{3}{10}x_2 - \frac{4}{10}x_3 = \frac{3}{2} - \frac{3}{10} \times (-3) - \frac{4}{10} \times 1 = 2$$

Therefore, the required solution by Crout's method (LU decomposition method) is

$$x_1 = 2, \ x_2 = -3, \ x_3 = 1.$$

Applications

Solving Linear Equations

Given a system of linear equations in matrix form

$$Ax = b,$$

we want to solve the equation for x, given A and b. Suppose we have already obtained the LUP decomposition of A such that $PA = LU$, so $LUx = Pb$.

In this case the solution is done in two logical steps:

1. First, we solve the equation $Ly = Pb$ for y.

2. Second, we solve the equation $Ux = y$ for x.

Note that in both cases we are dealing with triangular matrices (L and U), which can be solved directly by forward and backward substitution without using the Gaussian elimination process (however we do need this process or equivalent to compute the LU decomposition itself).

The above procedure can be repeatedly applied to solve the equation multiple times for different b. In this case it is faster (and more convenient) to do an LU decomposition of the matrix A once and then solve the triangular matrices for the different b, rather than using Gaussian elimination each time. The matrices L and U could be thought to have "encoded" the Gaussian elimination process.

The cost of solving a system of linear equations is approximately $\frac{2}{3}n^3$ floating-point operations if the matrix A has size n. This makes it twice as fast as algorithms based on the QR decomposition, which costs about $\frac{4}{3}n^3$ floating-point operations when Householder reflections are used. For this reason, the LU decomposition is usually preferred.

Inverting a Matrix

When solving systems of equations, b is usually treated as a vector with a length equal to the height of matrix A. In matrix inversion however, instead of vector b, we have matrix B, where B is an n-by-p matrix, so that we are trying to find a matrix X (also a n-by-p matrix):

$$AX = LUX = B.$$

We can use the same algorithm presented earlier to solve for each column of matrix X. Now suppose that B is the identity matrix of size n. It would follow that the result X must be the inverse of A.

Computing the Determinant

Given the LUP decomposition $A = P^{-1}LU$ of a square matrix A, the determinant of A can be computed straightforwardly as

$$\det(A) = \det(P^{-1})\det(L)\det(U) = (-1)^S \left(\prod_{i=1}^{n} l_{ii} \right) \left(\prod_{i=1}^{n} u_{ii} \right).$$

The second equation follows from the fact that the determinant of a triangular matrix is simply the product of its diagonal entries, and that the determinant of a permutation matrix is equal to $(-1)^S$ where S is the number of row exchanges in the decomposition.

In the case of LU decomposition with full pivoting, $\det(A)$ also equals the right-hand side of the above equation, if we let S be the total number of row and column exchanges.

The same method readily applies to LU decomposition by setting P equal to the identity matrix.

Cholesky Decomposition

The Cholesky method is used to solve a system of linear equations Ax = b if the coefficient matrix A is symmetric and positive definite. This method is also known as square-root method. Let A be a symmetric matrix, then A can be written as a product of lower triangular matrix and its transpose. That is,

$$A = LL^t,$$

where $L = [l_{ij}]$, $l_{ij} = 0$, $i < j$, a lower triangular matrix, L^T is the transpose of the matrix L.

Again, the matrix A can be written as

$$A = UU^t,$$

where U is an upper triangular matrix.

Using $(A = LL^t)$, the equation Ax = b becomes

$$LL^t x = b.$$

Let $\quad L^t x = z,$

Then $\quad Lz = b.$

Using forward substitution one can easily solve the equation to obtained the vector z. Then by solving the equation $L^t x = z$, using back substitution, we obtained the vector x.

Alternately, the values of z and then x can be determined from the following equations.

$$z = L^{-1}b \quad \text{and} \quad x = (L^t)^{-1}z = (L^{-1})^t z.$$

As an intermediate result the inverse of A can be determined from the following equation.

$$A^{-1} = (L^{-1})^t \, L^{-1}.$$

From this, it is clear that the solution of the system of equations is completely depends on the matrix L. The procedure to compute the matrix L is discussed below.

Procedure to Determine L

Since $A = LL^t$, then

$$
=\begin{bmatrix} l_{11} & 0 & 0 & \cdots & 0 \\ l_{21} & l_{22} & 0 & \cdots & 0 \\ \cdots & \cdots & \cdots & \cdots & \cdots \\ l_{i1} & l_{i2} & l_{i3} & \cdots & 0 \\ \cdots & \cdots & \cdots & \cdots & \cdots \\ l_{n1} & l_{n2} & l_{n3} & \cdots & l_{nn} \end{bmatrix}
\begin{bmatrix} l_{11} & l_{21} & \cdots & l_{j1} & \cdots & l_{n1} \\ 0 & l_{22} & \cdots & l_{j2} & \cdots & l_{n2} \\ 0 & 0 & \cdots & l_{j3} & \cdots & l_{n3} \\ \cdots & \cdots & \cdots & \cdots & \cdots & \cdots \\ 0 & 0 & \cdots & 0 & \cdots & l_{nn} \end{bmatrix}
$$

$$= \begin{bmatrix} l_{11}^2 & l_{21}l_{11} & \cdots & l_{j_1}l_{11} & \cdots & l_{n1}l_{11} \\ l_{21}l_{11} & l_{21}^2 + l_{22}^2 & \cdots & l_{j_1}l_{21} + l_{j_2}l_{22} & \cdots & l_{n1}l_{21} + l_{n2}l_{22} \\ \cdots & \cdots & \cdots & \cdots & \cdots & \cdots \\ l_{i1}l_{11} & l_{21}l_{i1} + l_{22}l_{i2} & \cdots & l_{j_1}l_{i1} + \cdots + l_{jj}l_{nj} & \cdots & l_{n1}l_{i1} + \cdots + l_{ni}l_{ii} \\ \cdots & \cdots & \cdots & \cdots & \cdots & \cdots \\ l_{n1}l_{11} & l_{21}l_{n1} + l_{22}l_{n2} & \cdots & l_{j_1}l_{n1} + \cdots + l_{jj}l_{nj} & \cdots & l_{n1}^2 + l_{n2}^2 + \cdots + l_{nn}^2 \end{bmatrix}.$$

By comparing both sides, we get the following system of equations.

$$l_{11} = (a_{11})^{1/2}$$

$$l_{i1}^2 + l_{i2}^2 + \cdots + l_{ii}^2 = a_{ii} \ \text{ or } \ l_{ii} = \left(a_{ii} - \sum_{j=1}^{i-1} l_{ij} \right)^{1/2}, i = 2,3,\ldots,n$$

$$l_{i1} = a_{i1} / l_{11}, i = 2,3,\ldots,n$$

$$l_{i1}l_{j1} + l_{i2}l_{j2} + \cdots + l_{ij}l_{jj} = a_{ij}$$

$$\text{or}, l_{ij} = \frac{1}{l_{ij}}\left(a_{ij} - \sum_{k=1}^{j-1} l_{jk}l_{ik} \right)^{1/2}, \ \text{ for } i = j+1, j+2, \ldots, n$$

$$l_{ij} = 0, i < j.$$

Note that this system of equations gives the values of l_{ij}.

Similarly, the elements of the matrix U for the system of equations $A = LL^t$, are given by,

$$u_{nn} = (a_{nn})^{1/2}$$

$$u_{in} = a_{in} / u_{nn}, i = 1, 2, \ldots, n-1$$

$$u_{ij} = \frac{1}{u_{jj}}(a_{ij} - \sum_{k=j+1}^{n} u_{ik}u_{jk}),$$

for $i = n-2, n-3, \ldots 1; \ j = i+1, i+2, \ldots, n-1$

$$u_{ii} = (a_{ii} - \sum_{k=i+1}^{n} u_{ik}^2)^{1/2}, i = n-1, n-2, \ldots, 1$$

$$u_{ij} = 0, i > j.$$

This method is illustrated by considering the following example,

Example: Solve the following system of equations by Cholesky method.

$$4x_1 + 2x_2 + 6x_3 = 16$$
$$2x_1 + 82x_2 + 39x_3 = 206$$
$$6x_1 + 39x_2 + 26x_3 = 113.$$

Solution: The given system of equations can be written as,

Ax = b where $x = (x_1, x_2 \ x_3)^t$, $b = (16, 206, 113)^t$, and $A = \begin{bmatrix} 4 & 2 & 6 \\ 2 & 82 & 39 \\ 6 & 39 & 26 \end{bmatrix}$.

Note that the coefficient matrix A is symmetric and positive definite and hence it can be written as $LL^t = A$.

$$\text{Let } L = \begin{bmatrix} l_{11} & 0 & 0 \\ l_{21} & l_{22} & 0 \\ l_{31} & l_{31} & l_{33} \end{bmatrix}.$$

Therefore, $LL^t = \begin{bmatrix} l_{11}^2 & l_{11}l_{21} & l_{11}l_{31} \\ l_{21}l_{11} & l_{21}^2 + l_{22}^2 & l_{21}l_{31} + l_{22}l_{32} \\ l_{31}l_{11} & l_{31}l_{21} + l_{32}l_{22} & l_{31}^2 + l_{33}^2 + l_{33}^2 \end{bmatrix} = \begin{bmatrix} 4 & 2 & 6 \\ 2 & 82 & 39 \\ 6 & 39 & 26 \end{bmatrix}.$

Comparing both sides, we get the following system of equations,

$$l_{11}^2 = 4 \text{ or } l_{11} = 2$$
$$l_{11}l_{21} = 2 \text{ or } l_{21} = 1$$
$$l_{11}l_{31} = 6 \text{ or } l_{21} = 3$$
$$l_{21}^2 + l_{22}^2 = 82 \text{ or } l_{22} = (82 - 1)^{1/2} = 9$$

$$l_{31}l_{21} + l_{32}l_{22} = 39 \text{ or } l_{32} = \frac{1}{l_{222}}(39 - l_{31}l_{21}) = 4$$

$$l_{31}^2 + l_{32}^2 + l_{33}^2 = 26 \text{ or } l_{33} = (26 - l_{31}^2 - l_{32}^2)^{1/2} = 1.$$

$$\text{Therefore, } L = \begin{bmatrix} 2 & 0 & 0 \\ 1 & 9 & 0 \\ 3 & 4 & 1 \end{bmatrix}.$$

Now, the system of equations Lz = b becomes,

$$2z_1 = 16$$
$$z_1 + 9z_2 = 206$$
$$3z_1 + 4z_2 + z_3 = 113.$$

The solution of these equations is,

$$z_1 = 8.0, \ z_2 = 22.0, \ z_3 = 1.0.$$

Now, from the equation $L^t x = z$.

$$\begin{bmatrix} 2.0 & 1.0 & 3.0 \\ 0.0 & 9.0 & 4.0 \\ 0.0 & 0.0 & 1.0 \end{bmatrix} \begin{bmatrix} x_1 \\ x_2 \\ x_3 \end{bmatrix} = \begin{bmatrix} 8.0 \\ 22.0 \\ 1.0 \end{bmatrix}.$$

In explicit form the equations are

$$2x_1 + x_2 \ 3x_3 = 8$$
$$9x_2 + 4x_3 = 22$$
$$x_3 = 1.$$

Solution of these equations is $x_3 = 1.0$, $x_2 = 2.0$, $x_1 = 1.5$.

Hence, the solution is $x_1 = 1.5$, $x_2 = 2.0$, $x_3 = 1.0$. This is the exact solution of the given system of equations.

Cholesky Example

For $A = \begin{pmatrix} 16 & 4 & 4 & -4 \\ 4 & 10 & 4 & 2 \\ 4 & 4 & 6 & -2 \\ -4 & 2 & -2 & 4 \end{pmatrix}$ and $b = \begin{pmatrix} 32 \\ 26 \\ 20 \\ -6 \end{pmatrix}$ construct a Cholesky decompostion and solve $Ax = b$.

Solution:

$$k = 1 : L_1 = \sqrt{16} = 4$$

$k = 2 : L_2 = 4, a_2 = 4, a_{22} = 10,$ Solve the 1×1 system $L_1 \ell_2 = a_2$ or $4\ell_2 = 4$ so $\ell_2 = 1$.
$\ell_{22} = \sqrt{10 - 1} = 3$. Therefore $L_2 = \begin{pmatrix} 4 & 0 \\ 1 & 3 \end{pmatrix}$.

$k = 3 : L_2 = \begin{pmatrix} 4 & 0 \\ 1 & 3 \end{pmatrix}, a_3 = \begin{pmatrix} 4 \\ 4 \end{pmatrix}, a_{33} = 6.$ Solve the 2×2 system

$L_2 \ell_3 = a_3$ or $\begin{pmatrix} 4 & 0 \\ 1 & 3 \end{pmatrix} \ell_3 = \begin{pmatrix} 4 \\ 4 \end{pmatrix}$ so $\ell_3 = \begin{pmatrix} 1 \\ 1 \end{pmatrix}$. $\ell_{33} = \sqrt{6 - (1 \ \ 1) \begin{pmatrix} 1 \\ 1 \end{pmatrix}} = 2.$

Therefore,

$$L_3 = \begin{pmatrix} 4 & 0 & 0 \\ 1 & 3 & 0 \\ 1 & 1 & 2 \end{pmatrix}$$

$k = 4 : L_3 = \begin{pmatrix} 4 & 0 & 0 \\ 1 & 3 & 0 \\ 1 & 1 & 2 \end{pmatrix}, a_4 \begin{pmatrix} -4 \\ 2 \\ -2 \end{pmatrix}, a_{44} = 4.$ Solve the 3×3 system

$L_3 \ell_4 = a_4$ or $\begin{pmatrix} 4 & 0 & 0 \\ 1 & 3 & 0 \\ 1 & 1 & 2 \end{pmatrix} \ell_4 = \begin{pmatrix} -4 \\ 2 \\ -2 \end{pmatrix}$ so $\ell_4 = \begin{pmatrix} -1 \\ 1 \\ -1 \end{pmatrix}$. $\ell_{44} = \sqrt{4 - (-1 \ \ 1 \ \ -1) \begin{pmatrix} -1 \\ 1 \\ -1 \end{pmatrix}} = 1.$

Therefore $L = L_4 = \begin{pmatrix} 4 & 0 & 0 & 0 \\ 1 & 3 & 0 & 0 \\ 1 & 1 & 2 & 0 \\ -1 & 1 & -1 & 1 \end{pmatrix}$

To solve $Ax = b$ note that $LL^T x = b$ which can be solved by solving $Lc = b$ and then $L^T x = c$.

$$Lc = b \Rightarrow c = \begin{pmatrix} 8 \\ 6 \\ 3 \\ -1 \end{pmatrix} \text{ and } L^T x = c \Rightarrow x = \begin{pmatrix} 1 \\ 2 \\ 1 \\ -1 \end{pmatrix}.$$

There are various methods for calculating the Cholesky decomposition. The computational complexity of commonly used algorithms is $O(n^3)$ in general. The algorithms described below all involve about $n^3/3$ FLOPs ($n^3/6$ multiplications and the same number of additions), where n is the size of the matrix A. Hence, they have half the cost of the LU decomposition, which uses $2n^3/3$ FLOPs.

Which of the algorithms below is faster depends on the details of the implementation. Generally, the first algorithm will be slightly slower because it accesses the data in a less regular manner.

The Cholesky Algorithm

The Cholesky algorithm, used to calculate the decomposition matrix L, is a modified version of Gaussian elimination.

The recursive algorithm starts with $i := 1$ and

$$A^{(1)} := A.$$

At step i, the matrix $A^{(i)}$ has the following form:

$$A^{(i)} = \begin{pmatrix} I_{i-1} & 0 & 0 \\ 0 & a_{i,i} & b_i^* \\ 0 & b_i & B^{(i)} \end{pmatrix},$$

where I_{i-1} denotes the identity matrix of dimension $i - 1$.

If we now define the matrix L_i by

$$L_i := \begin{pmatrix} I_{i-1} & O & O \\ O & \sqrt{a_{i,i}} & O \\ O & \dfrac{1}{\sqrt{a_{i,i}}} b_i & I_{n-i} \end{pmatrix},$$

then we can write $A^{(i)}$ as

$$A^{(i)} = L_i A^{(i+1)} L_i^*$$

Where

$$A^{(i+1)} = \begin{pmatrix} I_{i-1} & O & O \\ O & 1 & O \\ O & O & B^{(i)} - \dfrac{1}{a_{i,i}} b_i b_i^* \end{pmatrix}.$$

Note that $b_i\, b_i^*$ is an outer product, therefore this algorithm is called the *outer-product version* in (Golub & Van Loan).

We repeat this for i from 1 to n. After n steps, we get $A^{(n+1)} = I$. Hence, the lower triangular matrix L we are looking for is calculated as

$$L := L_1 L_2 \dots L_n.$$

The Cholesky–Banachiewicz and Cholesky–Crout Algorithms

If we write out the equation

$$A = LL^T = \begin{pmatrix} L_{11} & O & O \\ L_{21} & L_{22} & O \\ L_{31} & L_{32} & L_{33} \end{pmatrix} \begin{pmatrix} L_{11} & L_{21} & L_{31} \\ O & L_{22} & L_{32} \\ O & O & L_{33} \end{pmatrix}$$

$$= \begin{pmatrix} L_{11}^2 & & (\text{symmetric}) \\ L_{21}L_{11} & L_{21}^2 + L_{22}^2 & \\ L_{31}L_{11} & L_{31}L_{21} + L_{32}L_{22} & L_{31}^2 + L_{32}^2 + L_{33}^2 \end{pmatrix},$$

we obtain the following:

$$L = \begin{pmatrix} \sqrt{A_{11}} & O & O \\ A_{21}/L_{11} & \sqrt{A_{22} - L_{21}^2} & O \\ A_{31}/L_{11} & \left(A_{32} - L_{31}L_{21}\right)/L_{22} & \sqrt{A_{33} - L_{31}^2 - L_{32}^2} \end{pmatrix}$$

and therefore the following formulae for the entries of L:

$$L_{j,j} = \sqrt{A_{j,j} - \sum_{k=1}^{j-1} L_{j,k} L_{j,k}^2},$$

$$L_{i,j} = \frac{1}{L_{j,j}} \left(A_{i,j} - \sum_{k=1}^{j-1} L_{i,k} L_{j,k}^* \right) \quad \text{for } i > j.$$

The expression under the square root is always positive if A is real and positive-definite.

For complex Hermitian matrix, the following formula applies:

$$L_{j,j} = \sqrt{A_{j,j} - \sum_{k=1}^{j-1} L_{j,k} L_{j,k}^*},$$

$$L_{i,j} = \frac{1}{L_{j,j}} \left(A_{i,j} - \sum_{k=1}^{j-1} L_{i,k} L_{j,k}^* \right) \quad \text{for } i > j.$$

So we can compute the (i, j) entry if we know the entries to the left and above. The computation is usually arranged in either of the following orders:

- The Cholesky–Banachiewicz algorithm starts from the upper left corner of the matrix L and proceeds to calculate the matrix row by row.

- The Cholesky–Crout algorithm starts from the upper left corner of the matrix L and proceeds to calculate the matrix column by column.

Either pattern of access allows the entire computation to be performed in-place if desired.

Stability of the Computation

Suppose that we want to solve a well-conditioned system of linear equations. If the LU decomposition is used, then the algorithm is unstable unless we use some sort of pivoting strategy. In the latter case, the error depends on the so-called growth factor of the matrix, which is usually (but not always) small.

Now, suppose that the Cholesky decomposition is applicable. As mentioned above, the algorithm will be twice as fast. Furthermore, no pivoting is necessary, and the error will always be small. Specifically, if we want to solve Ax = b, and y denotes the computed solution, then y solves the perturbed system (A + E)y = b, where

$$\| E \|_2 \leq c_n \varepsilon \| A \|_2 .$$

Here $||\cdot||_2$ is the matrix 2-norm, c_n is a small constant depending on n, and ε denotes the unit round-off.

One concern with the Cholesky decomposition to be aware of is the use of square roots. If the matrix being factorized is positive definite as required, the numbers under the square roots are always positive *in exact arithmetic*. Unfortunately, the numbers can become negative because of round-off errors, in which case the algorithm cannot continue. However, this can only happen if the matrix is very ill-conditioned. One way to address this is to add a diagonal correction matrix to the matrix being decomposed in an attempt to promote the positive-definiteness. While this might lessen the accuracy of the decomposition, it can be very favorable for other reasons; for example, when performing Newton's method in optimization, adding a diagonal matrix can improve stability when far from the optimum.

LDL Decomposition

An alternative form, eliminating the need to take square roots, is the symmetric indefinite factorization,

$$A = LDL^T = \begin{pmatrix} 1 & 0 & 0 \\ L_{21} & 1 & 0 \\ L_{31} & L_{32} & 1 \end{pmatrix} \begin{pmatrix} D_1 & 0 & 0 \\ 0 & D_2 & 0 \\ 0 & 0 & D_3 \end{pmatrix} \begin{pmatrix} 1 & L_{21} & L_{31} \\ 0 & 1 & L_{32} \\ 0 & 0 & 1 \end{pmatrix}$$

$$= \begin{pmatrix} D_1 & & \text{(symmetric)} \\ L_{21}D_1 & L_{21}^2 D_1 + D_2 & \\ L_{31}D_1 & L_{31}L_{21}D_1 + L_{32}D_2 & L_{31}^2 D_1 + L_{32}^2 D_2 + D_3 \cdot \end{pmatrix}.$$

If A is real, the following recursive relations apply for the entries of D and L:

$$D_j = A_{jj} - \sum_{k=1}^{j-1} L_{jk}^2 D_k,$$

$$L_{ij} = \frac{1}{D_j} \left(A_{ij} - \sum_{k=1}^{j-1} L_{ik} L_{jk} D_k \right) \quad \text{for } i > j.$$

For complex Hermitian matrix A, the following formula applies:

$$D_j = A_{jj} - \sum_{k=1}^{j-1} L_{jk} L_{jk}^* D_k,$$

$$L_{ij} = \frac{1}{D_j} \left(A_{ij} - \sum_{k=1}^{j-1} L_{ik} L_{jk}^* D_k \right) \quad \text{for } i > j.$$

Again, the pattern of access allows the entire computation to be performed in-place if desired.

Block Variant

When used on indefinite matrices, the LDL* factorization is known to be unstable without careful pivoting; specifically, the elements of the factorization can grow arbitrarily. A possible improvement is to perform the factorization on block sub-matrices, commonly 2×2:

$$A = LDL^T = \begin{pmatrix} I & O & O \\ L_{21} & I & O \\ L_{31} & L_{32} & I \end{pmatrix} \begin{pmatrix} D_1 & O & O \\ O & D_2 & O \\ O & O & D_3 \end{pmatrix} \begin{pmatrix} I & L_{21}^T & L_{31}^T \\ O & I & L_{32}^T \\ O & O & I \end{pmatrix}$$

$$= \begin{pmatrix} D_1 & & \text{(symmetric)} \\ L_{21}D_1 & L_{21}D_1L_{21}^T + D_2 & \\ L_{31}D_1 & L_{31}D_1L_{21}^T + L_{32}D_2 & L_{31}D_1L_{31}^T + L_{32}D_2L_{32}^T + D_3 \end{pmatrix},$$

where every element in the matrices above is a square submatrix. From this, these analogous recursive relations follow:

$$D_j = A_{jj} - \sum_{k=1}^{j-1} L_{jk}D_kL_{jk}^T,$$

$$L_{ij} = \left(A_{ij} - \sum_{k=1}^{j-1} L_{ik}D_kL_{jk}^T \right) D_j^{-1}.$$

This involves matrix products and explicit inversion, thus limiting the practical block size.

Updating the Decomposition

A task that often arises in practice is that one needs to update a Cholesky decomposition. In more details, one has already computed the Cholesky decomposition $A = LL^*$ of some matrix A, then one changes the matrix A in some way into another matrix, say $\tilde{A}$, and one wants to compute the Cholesky decomposition of the updated matrix: $\tilde{A} = \tilde{L}\tilde{L}^*$. The question is now whether one can use the Cholesky decomposition of A that was computed before to compute the Cholesky decomposition of $\tilde{A}$.

Rank-one Update

The specific case, where the updated matrix $\tilde{A}$. is related to the matrix A by $\tilde{A} = A + xx^*$, is known as a rank-one update.

Here is a little function written in Matlab syntax that realizes a rank-one update:

```
function [L] = cholupdate(L, x)
```

```
n = length(x);
for k = 1:n
    r = sqrt(L(k, k)^2 + x(k)^2);
    c = r / L(k, k);
    s = x(k) / L(k, k);
    L(k, k) = r;
    if k < n
        L((k+1):n, k) = (L((k+1):n, k) + s * x((k+1):n)) / c;
        x((k+1):n) = c * x((k+1):n) - s * L((k+1):n, k);
    end
    end
end
```

Rank-one Downdate

A *rank-one downdate* is similar to a rank-one update, except that the addition is replaced by subtraction: $\widetilde{A} = A - xx^*$. This only works if the new matrix $\widetilde{A}$ is still positive definite.

The code for the rank-one update shown above can easily be adapted to do a rank-one downdate: one merely needs to replace the two additions in the assignment to r and L((k+1):n,k) by subtractions.

Adding and Removing Rows and Columns

If we have a symmetric and positive definite matrix A represented in block form as,

$$A \;\; = \begin{pmatrix} A_{11} & A_{13} \\ A_{13}^T & A_{33} \end{pmatrix}$$

And its upper Cholesky factor,

$$L = \begin{pmatrix} L_{11} & L_{13} \\ 0 & L_{33} \end{pmatrix}$$

Then, for a new matrix $\widetilde{A}$ which is the same as A but with the insertion of new rows and columns,

$$\widetilde{A} = \begin{pmatrix} A_{11} & A_{12} & A_{13} \\ A_{12}^T & A_{22} & A_{23} \\ A_{13}^T & A_{23}^T & A_{33} \end{pmatrix}$$

we are interested in finding the Cholesky factorisation of $\widetilde{A}$, which we call $\widetilde{S}$, without directly computing the entire decomposition.

$$\widetilde{S} = \begin{pmatrix} S_{11} & S_{12} & S_{13} \\ 0 & S_{22} & S_{23} \\ 0 & 0 & S_{33} \end{pmatrix}$$

Writing $A \backslash b$ for the solution of $Ax = b$, which can be found easily for triangular matrices, and (M) for the Cholesky decomposition of M, the following relations can be found;

$$S_{11} = L_{11}$$
$$S_{12} = L_{11}^T \backslash A_{12} S_{13} = L_{13}$$
$$S_{22} = \text{chol}(A_{22} - S_{12}^T S_{12})$$
$$S_{23} = S_{22}^T \backslash (A_{23} - S_{12}^T S_{13})$$
$$S_{33} = \text{chol}(L_{33}^T L_{33} - S_{23}^T S_{23})$$

These formulae may be used to determine the Cholesky factor after the insertion of rows or columns in any position, if we set the row and column dimensions appropriately (including to zero). The inverse problem, when we have,

$$\widetilde{A} = \begin{pmatrix} A_{11} & A_{12} & A_{13} \\ A_{12}^T & A_{22} & A_{23} \\ A_{13}^T & A_{23}^T & A_{33} \end{pmatrix}$$

with known Cholesky decomposition,

$$\widetilde{S} = \begin{pmatrix} S_{11} & S_{12} & S_{13} \\ 0 & S_{22} & S_{23} \\ 0 & 0 & S_{33} \end{pmatrix}$$

And we wish to determine the Cholesky factor,

$$L = \begin{pmatrix} L_{11} & L_{13} \\ 0 & L_{33} \end{pmatrix}$$

of the matrix A with rows and columns removed,

$$A = \begin{pmatrix} A_{11} & A_{13} \\ A_{13}^T & A_{33} \end{pmatrix}$$

yields the following rules,

$$L_{11} = S_{11}$$
$$L_{13} = S_{13}$$
$$L_{33} = \text{chol}(S_{33}^T S_{33} + S_{23}^T S_{23})$$

Notice that the equations above that involve finding the Cholesky decomposition of a new matrix are all of the form $\widetilde{A} = A \pm xx^*$, which allows them to be efficiently calculated using the update and downdate procedures.

Proof for Positive Semi-definite Matrices

The above algorithms show that every positive definite matrix A has a Cholesky decomposition. This result can be extended to the positive semi-definite case by a limiting argument. The argument is not fully constructive, i.e., it gives no explicit numerical algorithms for computing Cholesky factors.

If A is an $n \times n$ positive semi-definite matrix, then the sequence $(A_k)_k := \left(A + \dfrac{1}{k}I_n\right)_k$

consists of positive definite matrices. (This is an immediate consequence of, for example, the spectral mapping theorem for the polynomial functional calculus.) Also,

$$A_k \to A \quad \text{for} \quad k \to \infty$$

in operator norm. From the positive definite case, each A_k has Cholesky decomposition $A_k = L_k L_k^*$. By property of the operator norm,

$$\| L_k \|^2 \geq \| L_k L_k^* \| = \| A_k \|.$$

So $(L_k)_k$ is a bounded set in the Banach space of operators, therefore relatively compact (because the underlying vector space is finite-dimensional). Consequently, it has a convergent subsequence, also denoted by $(L_k)_k$, with limit L. It can be easily checked that this L has the desired properties, i.e. $A = LL^*$, and L is lower triangular with non-negative diagonal entries: for all x and y,

$$\langle Ax, y \rangle = \langle \lim A_k x, y \rangle = \langle \lim L_k L_k^* x, y \rangle = \langle LL^* x, y \rangle.$$

Therefore, $A = LL^*$. because the underlying vector space is finite-dimensional, all topologies on the space of operators are equivalent. So $(L_k)_k$ tends to L in norm means $(L_k)_k$ tends to L entrywise. This in turn implies that, since each L_k is lower triangular with non-negative diagonal entries, L is also.

Generalization

The Cholesky factorization can be generalized to (not necessarily finite) matrices with operator entries. Let $\{\mathcal{H}_n\}$ be a sequence of Hilbert spaces. Consider the operator matrix,

$$A = \begin{bmatrix} A_{11} & A_{12} & A_{13} \\ A_{12}^* & A_{22} & A_{23} \\ A_{13}^* & A_{23}^* & A_{33} \\ & & & \ddots \end{bmatrix}$$

acting on the direct sum,

$$\mathcal{H} = \oplus_n \mathcal{H}_n,$$

where each,

$$A_{ij} : \mathcal{H}_j \to \mathcal{H}_i$$

is a bounded operator. If A is positive (semidefinite) in the sense that for all finite k and for any,

$$h \in \oplus_{n=1}^k \mathcal{H}_k,$$

we have $\langle h, Ah \rangle \geq 0,$ then there exists a lower triangular operator matrix L such that A = LL*. One can also take the diagonal entries of L to be positive.

Applications

The Cholesky decomposition is mainly used for the numerical solution of linear equations $Ax = b$. If A is symmetric and positive definite, then we can solve $Ax = b$ by first computing the Cholesky decomposition $A = LL^*$, then solving $Ly = b$ for y by forward substitution, and finally solving $L^*x = y$ for x by back substitution.

An alternative way to eliminate taking square roots in the LL^* decomposition is to compute the Cholesky decomposition $A = LDL^*$, then solving $Ly = b$ for y, and finally solving $DL^*x = y$.

For linear systems that can be put into symmetric form, the Cholesky decomposition (or its LDL variant) is the method of choice, for superior efficiency and numerical stability. Compared to the LU decomposition, it is roughly twice as efficient.

Linear Least Squares

Systems of the form $Ax = b$ with A symmetric and positive definite arise quite often in applications. For instance, the normal equations in linear least squares problems are of this form. It may also happen that matrix A comes from an energy functional, which must be positive from physical considerations; this happens frequently in the numerical solution of partial differential equations.

Non-linear Optimization

Non-linear multi-variate functions may be minimized over their parameters using variants of Newton's method called *quasi-Newton* methods. At each iteration, the search takes a step s defined by solving Hs = −g for s, where s is the step, g is the *gradient* vector of the function's partial first derivatives with respect to the parameters, and H is an approximation to the Hessian matrix of partial second derivatives formed by repeated rank-1 updates at each iteration. Two well-known update formulae are called Davidon–Fletcher–Powell (DFP) and Broyden–Fletcher–Goldfarb–Shanno (BFGS). Loss of the positive-definite condition through round-off error is avoided if rather than updating an approximation to the inverse of the Hessian, one updates the Cholesky decomposition of an approximation of the Hessian matrix itself.

Monte Carlo Simulation

The Cholesky decomposition is commonly used in the Monte Carlo method for simulating systems with multiple correlated variables. The correlation matrix is decomposed, to give the lower-triangular L. Applying this to a vector of uncorrelated samples u produces a sample vector Lu with the covariance properties of the system being modeled.

For a simplified example that shows the economy one gets from the Cholesky decomposition, say one needs to generate two correlated normal variables x_1 and x_2 with given correlation coefficient ρ. All one needs to do is to generate two uncorrelated Gaussian random variables z_1 and z_2. We set $x_1 = z_1$ and $x_2 = \rho z_1 + \sqrt{1 - \rho^2}\, z_2$.

Kalman Filters

Unscented Kalman filters commonly use the Cholesky decomposition to choose a set of so-called sigma points. The Kalman filter tracks the average state of a system as a vector x of length N and covariance as an $N \times N$ matrix P. The matrix P is always positive semi-definite and can be decomposed into LL^T. The columns of L can be added and subtracted from the mean x to form a set of $2N$ vectors called *sigma points*. These sigma points completely capture the mean and covariance of the system state.

Matrix Inversion

The explicit inverse of a Hermitian matrix can be computed by Cholesky decomposition, in a manner similar to solving linear systems, using n^3 operations ($\frac{1}{2}n^3$ multiplications). The entire inversion can even be efficiently performed in-place.

A non-Hermitian matrix B can also be inverted using the following identity, where BB* will always be Hermitian:

$$B^{-1} = B^*(BB^*)^{-1}.$$

QR Decomposition

Let A be an $m \times n$ matrix with full column rank. The QR factorization of A is a decomposition $A = QR$, where Q is an $m \times m$ orthogonal matrix and R is an $m \times n$ upper triangular matrix. There are three ways to compute this decomposition:

1. Using Householder matrices

2. Using Givens rotations, also known as Jacobi rotations

3. A third, less frequently used approach is the Gram-Schmidt orthogonalization.

Givens Rotations

We illustrate the process in the case where A is a 2×2 matrix. In Gaussian elimination, we compute $L^{-1} A = U$ where L^{-1} is unit lower triangular and U is upper triangular. Specifically,

$$\begin{bmatrix} 1 & 0 \\ m_{21} & 1 \end{bmatrix} \begin{bmatrix} a_{11} & a_{12} \\ a_{21} & a_{22} \end{bmatrix} = \begin{bmatrix} a_{11}^{(2)} & a_{12}^{(2)} \\ 0 & a_{22}^{(2)} \end{bmatrix}, \quad m_{21} = -\frac{a_{21}}{a_{11}}.$$

By contrast, the QR decomposition computes $Q^T A = R$, or

$$\begin{bmatrix} \gamma & -\sigma \\ \sigma & \gamma \end{bmatrix}^T \begin{bmatrix} a_{11} & a_{12} \\ a_{21} & a_{22} \end{bmatrix} = \begin{bmatrix} r_{11} & r_{12} \\ 0 & r_{22} \end{bmatrix},$$

where $\gamma^2 + \sigma^2 = 1$.

From the relationship $-\sigma a_{11} + \gamma a_{21} = 0$ we obtain,

$$\gamma a_{21} = \sigma a_{11}$$

$$\gamma^2 a_{21}^2 = \sigma^2 a_{21}^2 = (1 - \gamma^2) a_{11}^2$$

which yields,

$$\gamma = \pm \frac{a_{11}}{\sqrt{a_{21}^2 + a_{11}^2}}.$$

It is conventional to choose the + sign. Then, we obtain,

$$\sigma^2 = 1 - \gamma^2 = 1 - \frac{a_{11}^2}{a_{21}^2 + a_{11}^2} = \frac{a_{21}^2}{a_{21}^2 + a_{11}^2},$$

or

$$\sigma = \pm \frac{a_{21}}{\sqrt{a_{21}^2 + a_{11}^2}}.$$

Again, we choose the + sign. As a result, we have,

$$r_{11} = a_{11} \frac{a_{11}}{\sqrt{a_{21}^2 + a_{11}^2}} + a_{21} \frac{a_{21}}{\sqrt{a_{21}^2 + a_{11}^2}} = \sqrt{a_{21}^2 + a_{11}^2}.$$

The matrix,

$$Q = \begin{bmatrix} \gamma & -\sigma \\ \sigma & \gamma \end{bmatrix}^T$$

is called a Givens rotation. It is called a rotation because it is orthogonal, and therefore length preserving, and also because there is an angle θ such that $\sin \theta = \sigma$ and $\cos \theta = \gamma$, and its effect is to rotate a vector clockwise through the angle θ. In particular,

$$\begin{bmatrix} \gamma & -\sigma \\ \sigma & \gamma \end{bmatrix}^T \begin{bmatrix} \alpha \\ \beta \end{bmatrix} = \begin{bmatrix} \rho \\ 0 \end{bmatrix}$$

where $\rho = \sqrt{\alpha^2 + \beta^2}, \alpha = \rho \cos \theta$ and $\beta = \rho \sin \theta$. It is easy to verify that the product of two rotations is itself a rotation. Now, in the case where A is an $n \times n$ matrix, suppose that we have the vector:

So, in order to transform A into an upper triangular matrix R, we can find a product of rotations Q such that $Q^T A = R$, It is easy to see that $O(n^2)$ rotations are required. Each rotation takes $O(n)$ operations, so the entire process of computing the QR factorization requires $O(n^3)$ operations.

It is important to note that the straightforward approach to computing the entries γ and σ of the Givens rotation,

$$\gamma = \frac{\alpha}{\sqrt{\alpha^2 + \beta^2}}, \sigma = \frac{\beta}{\sqrt{\alpha + 2 + \beta^2}},$$

is not always advisable, because in floating-point arithmetic, the computation of $\sqrt{\alpha^2 + \beta^2}$ could overflow. To get around this problem, suppose that $|\beta| \geq |\alpha|$. Then, we can instead compute,

$$\tau = \frac{\alpha}{\beta}, \ \sigma = \frac{1}{\sqrt{1 + \tau^2}}, \ \gamma = \sigma\tau,$$

which is guaranteed not to overflow since the only number that is squared is less than one in magnitude. On the other hand, if $|\alpha| \geq |\beta|$, then we compute,

$$\tau = \frac{\beta}{\alpha}, \ \gamma = \frac{1}{\sqrt{1 + \tau^2}}, \ \sigma = \gamma\tau,$$

Now, we describe the entire algorithm for computing the QR factorization using Givens rotations. Let $[c, s]$ = givens (a, b) be a Matlab-style function that computes c and s such that,

$$\begin{bmatrix} c & -s \\ s & c \end{bmatrix}^T \begin{bmatrix} a \\ b \end{bmatrix} = \begin{bmatrix} r \\ 0 \end{bmatrix}, \ r = \sqrt{a^2 + b^2}.$$

Then, let $G(i, j, c, s)^T$ be the Givens rotation matrix that rotates the ith and jth elements of a vector v clockwise by the angle θ such that $\cos\theta = c$ and $\sin\theta = s$, so that if $v_i = a$ and $v_j = b$, and $[[c,s]$ = givens (a, b), then in the updated vector u $G(i, j, c, s)^T v, u_i = r = \sqrt{a^2 + b^2}$ and $u_j = 0$. The QR factorization of an $m \times n$ matrix A is then computed as follows.

$Q = I$
$R = A$ *for* $j = 1 : n$ do
    for $i = m : -1 : j + 1$ do
    $[c,s]$ = givens $(r_{1-1,j}, r_{ij})$
      $R = G(i, j, c, s)^T R$
      $Q = QG(i, j, c, s)$
    end
  end

Note that the matrix Q is accumulated by column rotations of the identity matrix, because the matrix by which A is multiplied to reduce A to upper-triangular form, a product of row rotations, is Q^T.

Example: We use Givens rotations to compute the QR factorization of,

$$\begin{bmatrix} 0.8147 & 0.0975 & 0.1576 \\ 0.9058 & 0.2785 & 0.9706 \\ 0.1270 & 0.5469 & 0.9572 \\ 0.9134 & 0.9575 & 0.4854 \\ 0.6324 & 0.9649 & 0.8003 \end{bmatrix}.$$

First, we compute a Givens rotation that, when applied to a_{41} and a_{51}, zeros a_{51}:

$$\begin{bmatrix} 0.8222 & -0.5692 \\ 0.5692 & 0.8222 \end{bmatrix}^T \begin{bmatrix} 0.9134 \\ 0.6324 \end{bmatrix} = \begin{bmatrix} 1.1109 \\ 0 \end{bmatrix}.$$

Applying this rotation to rows 4 and 5 yields

$$\begin{bmatrix} 1 & 0 & 0 & 0 & 0 \\ 0 & 1 & 0 & 0 & 0 \\ 0 & 0 & 1 & 0 & 0 \\ 0 & 0 & 0 & 0.8222 & -0.5692 \\ 0 & 0 & 0 & 0.5692 & 0.8222 \end{bmatrix}^T \begin{bmatrix} 0.8147 & 0.0975 & 0.1576 \\ 0.9058 & 0.2785 & 0.9706 \\ 0.1270 & 0.5469 & 0.9572 \\ 0.9134 & 0.9575 & 0.4854 \\ 0.6324 & 0.9649 & 0.8003 \end{bmatrix} = \begin{bmatrix} 0.8147 & 0.0975 & 0.1576 \\ 0.9058 & 0.2785 & 0.9706 \\ 0.1270 & 0.5469 & 0.9572 \\ 1.1109 & 1.3365 & 0.8546 \\ 0 & 0.2483 & 0.3817 \end{bmatrix}$$

Next, we compute a Givens rotation that, when applied to a_{31} and a_{41}, zeros a_{41}:

$$\begin{bmatrix} 0.1136 & -0.9935 \\ 0.9935 & 0.1136 \end{bmatrix}^T \begin{bmatrix} 0.1270 \\ 0.1109 \end{bmatrix} = \begin{bmatrix} 1.1181 \\ 0 \end{bmatrix}.$$

Applying this rotation to rows 3 and 4 yields

$$\begin{bmatrix} 1 & 0 & 0 & 0 & 0 \\ 0 & 1 & 0 & 0 & 0 \\ 0 & 0 & 0.1136 & -0.9935 & 0 \\ 0 & 0 & 0.9935 & 0.1136 & 0 \\ 0 & 0 & 0 & 0 & 1 \end{bmatrix}^T \begin{bmatrix} 0.8147 & 0.0975 & 0.1576 \\ 0.9058 & 0.2785 & 0.9706 \\ 0.1270 & 0.5469 & 0.9572 \\ 1.1109 & 1.3365 & 0.8546 \\ 0 & 0.2483 & 0.3817 \end{bmatrix} = \begin{bmatrix} 0.8147 & 0.0975 & 0.1576 \\ 0.9058 & 0.2785 & 0.9706 \\ 1.1181 & 1.3899 & 0.9572 \\ 0 & -0.3916 & -0.8539 \\ 0 & 0.2483 & 0.3817 \end{bmatrix}$$

Next, we compute a Givens rotation that, when applied to a_{21} and a_{31}, zeros a_{31}:

$$\begin{bmatrix} 0.6295 & -0.7770 \\ 0.7770 & 0.6295 \end{bmatrix}^T \begin{bmatrix} 0.9058 \\ 1.1181 \end{bmatrix} = \begin{bmatrix} 1.4390 \\ 0 \end{bmatrix}.$$

Applying this rotation to rows 2 and 3 yields,

$$
\begin{bmatrix} 1 & 0 & 0 & 0 & 0 \\ 0 & 0.6295 & -0.7770 & 0 & 0 \\ 0 & 0.7770 & 0.6295 & 0 & 0 \\ 0 & 0 & 0 & 1 & 0 \\ 0 & 0 & 0 & 0 & 1 \end{bmatrix}^T
\begin{bmatrix} 0.8147 & 0.0975 & 0.1576 \\ 0.9058 & 0.2785 & 0.9706 \\ 1.1181 & 1.3899 & 0.9578 \\ 0 & -0.3916 & 0.8539 \\ 0 & 0.2483 & 0.3817 \end{bmatrix}
=
\begin{bmatrix} 0.8147 & 0.0975 & 0.1576 \\ 1.4390 & 1.2553 & 1.3552 \\ 0 & 0.6585 & -0.1513 \\ 0 & -0.3916 & -0.8539 \\ 0 & 0.2483 & 0.3817 \end{bmatrix}.
$$

To complete the first column, we compute a Givens rotation that, when applied to a_{11} and a_{21}, zeros a_{21}:

$$
\begin{bmatrix} 0.4927 & -0.8702 \\ 0.8702 & 0.4927 \end{bmatrix}^T
\begin{bmatrix} 0.8147 \\ 1.4390 \end{bmatrix}
=
\begin{bmatrix} 1.6536 \\ 0 \end{bmatrix}.
$$

Applying this rotation to rows 1 and 2 yields,

$$
\begin{bmatrix} 0.4927 & -0.8702 & 0 & 0 & 0 \\ 0.8702 & 0.4927 & 0 & 0 & 0 \\ 0 & 0 & 1 & 0 & 0 \\ 0 & 0 & 0 & 1 & 0 \\ 0 & 0 & 0 & 0 & 1 \end{bmatrix}^T
\begin{bmatrix} 0.8147 & 0.0975 & 0.1576 \\ 1.4390 & 1.2553 & 1.3552 \\ 0 & 0.6585 & -0.1513 \\ 0 & -0.3916 & -0.8539 \\ 0 & 0.2483 & 0.3817 \end{bmatrix}
=
\begin{bmatrix} 1.6536 & 1.1405 & 1.2539 \\ 0 & 0.5336 & 0.5305 \\ 0 & 0.6585 & -0.1513 \\ 0 & -0.3916 & -0.8539 \\ 0 & 0.2483 & 0.3817 \end{bmatrix}.
$$

Moving to the second column, we compute a Givens rotation that, when applied to a_{42} and a_{52}, zeros a_{52}:

$$
\begin{bmatrix} 0.8445 & 0.5355 \\ -0.5355 & 0.8445 \end{bmatrix}^T
\begin{bmatrix} -0.3916 \\ 0.2483 \end{bmatrix}
=
\begin{bmatrix} 0.4636 \\ 0 \end{bmatrix}.
$$

Applying this rotation to rows 4 and 5 yields,

$$
\begin{bmatrix} 1 & 0 & 0 & 0 & 0 \\ 0 & 1 & 0 & 0 & 0 \\ 0 & 0 & 1 & 0 & 0 \\ 0 & 0 & 0 & 0.8445 & 0.5355 \\ 0 & 0 & 0 & -0.5355 & 0.8445 \end{bmatrix}^T
\begin{bmatrix} 1.6536 & 1.1405 & 1.2569 \\ 0 & 0.5336 & 0.5305 \\ 0 & 0.6585 & -0.1513 \\ 0 & -0.3916 & -0.8539 \\ 0 & 0.2483 & 0.3817 \end{bmatrix}
=
\begin{bmatrix} 1.6536 & 1.1405 & 1.2539 \\ 0 & 0.5336 & 0.5305 \\ 0 & 0.6585 & -0.1513 \\ 0 & -0.4636 & -0.9256 \\ 0 & 0 & -0.1349 \end{bmatrix}.
$$

Next, we compute a Givens rotation that, when applied to a_{32} and a_{42}, zeros a_{42}:

$$
\begin{bmatrix} 0.8177 & 0.5757 \\ -0.5757 & 0.8177 \end{bmatrix}^T
\begin{bmatrix} 0.6585 \\ -0.4636 \end{bmatrix}
=
\begin{bmatrix} 0.8054 \\ 0 \end{bmatrix}.
$$

Applying this rotation to rows 3 and 4 yields,

$$
\begin{bmatrix} 1 & 0 & 0 & 0 & 0 \\ 0 & 1 & 0 & 0 & 0 \\ 0 & 0 & 0.8177 & 0.5757 & 0 \\ 0 & 0 & -0.5757 & 0.8177 & 0 \\ 0 & 0 & 0 & 0 & 1 \end{bmatrix}^{T}
\begin{bmatrix} 1.6536 & 1.1405 & 1.2569 \\ 0 & 0.5336 & 0.5305 \\ 0 & 0.6585 & -0.1513 \\ 0 & -0.4636 & -0.9256 \\ 0 & 0 & 0.1349 \end{bmatrix} =
\begin{bmatrix} 1.6536 & 1.1405 & 1.2569 \\ 0 & 0.5336 & 0.5305 \\ 0 & 0.8054 & 0.4091 \\ 0 & 0 & -0.8439 \\ 0 & 0 & -0.1349 \end{bmatrix}.
$$

Next, we compute a Givens rotation that, when applied to a_{22} and a_{32}, zeros a_{22}:

$$
\begin{bmatrix} 0.5523 & -0.8336 \\ 0.8336 & 0.5523 \end{bmatrix}^{T}
\begin{bmatrix} 0.5336 \\ 0.8054 \end{bmatrix} =
\begin{bmatrix} 0.9661 \\ 0 \end{bmatrix}.
$$

Applying this rotation to rows 3 and 4 yields,

$$
\begin{bmatrix} 1 & 0 & 0 & 0 & 0 \\ 0 & 0.5523 & -0.8336 & 0 & 0 \\ 0 & 0.8336 & 0.5523 & 0 & 0 \\ 0 & 0 & 0 & 1 & 0 \\ 0 & 0 & 0 & 0 & 1 \end{bmatrix}^{T}
\begin{bmatrix} 1.6536 & 1.1405 & 1.2569 \\ 0 & 0.5336 & 0.5305 \\ 0 & 0.8054 & 0.4090 \\ 0 & 0 & -0.8439 \\ 0 & 0 & -0.1349 \end{bmatrix} =
\begin{bmatrix} 1.6536 & 1.1405 & 1.2569 \\ 0 & 0.9661 & 0.6341 \\ 0 & 0 & -0.2163 \\ 0 & 0 & -0.8439 \\ 0 & 0 & -0.1349 \end{bmatrix}.
$$

Moving to the third column, we compute a Givens rotation that, when applied to a_{43} and a_{53}, zeros $_{53}$

$$
\begin{bmatrix} 0.9875 & -0.1579 \\ 0.1579 & 0.9875 \end{bmatrix}^{T}
\begin{bmatrix} -0.8439 \\ -0.1349 \end{bmatrix} =
\begin{bmatrix} 0.8546 \\ 0 \end{bmatrix}.
$$

Applying this rotation to rows 4 and 5 yields,

$$
\begin{bmatrix} 1 & 0 & 0 & 0 & 0 \\ 0 & 1 & 0 & 0 & 0 \\ 0 & 0 & 1 & 0 & 0 \\ 0 & 0 & 0 & 0.9875 & -0.1579 \\ 0 & 0 & 0 & 0.1579 & 0.9875 \end{bmatrix}^{T}
\begin{bmatrix} 1.6536 & 1.1405 & 1.2569 \\ 0 & 0.9661 & 0.6341 \\ 0 & 0 & -0.2163 \\ 0 & 0 & -0.8439 \\ 0 & 0 & -0.1349 \end{bmatrix} =
\begin{bmatrix} 1.6536 & 1.1405 & 1.2569 \\ 0 & 0.9661 & 0.6341 \\ 0 & 0 & -0.2163 \\ 0 & 0 & -0.8546 \\ 0 & 0 & 0 \end{bmatrix}.
$$

Finally, we compute a Givens rotation that, when applied to a_{33} and a_{43}, zeros a_{43}:

$$
\begin{bmatrix} 0.2453 & -0.9694 \\ 0.9694 & 0.2453 \end{bmatrix}^{T}
\begin{bmatrix} -0.2163 \\ -0.8546 \end{bmatrix} =
\begin{bmatrix} 0.8816 \\ 0 \end{bmatrix}.
$$

Applying this rotation to rows 3 and 4 yields,

$$
\begin{bmatrix}
1 & 0 & 0 & 0 & 0 \\
0 & 1 & 0 & 0 & 0 \\
0 & 0 & 0.2453 & -0.9694 & 0 \\
0 & 0 & 0.9694 & 0.2453 & 0 \\
0 & 0 & 0 & 0 & 1
\end{bmatrix}^{T}
\begin{bmatrix}
1.6536 & 1.1405 & 1.2569 \\
0 & 0.9661 & 0.6341 \\
0 & 0 & -0.2163 \\
0 & 0 & -0.8439 \\
0 & 0 & 0
\end{bmatrix} =
\begin{bmatrix}
1.6536 & 1.1405 & 1.2569 \\
0 & 0.9661 & 0.6341 \\
0 & 0 & -0.8816 \\
0 & 0 & 0 \\
0 & 0 & 0
\end{bmatrix} = .
$$

Applying the transpose of each Givens rotation, in order, to the columns of the identity matrix yields the matrix,

$$
Q =
\begin{bmatrix}
0.4927 & -0.4806 & 0.1780 & -0.7033 & 0 \\
0.5478 & -0.3583 & -0.5777 & 0.4825 & 0.0706 \\
0.0768 & 0.4754 & -0.6343 & -0.4317 & -0.4235 \\
0.5523 & 0.3391 & 0.4808 & 0.2769 & -0.5216 \\
0.3824 & 0.5473 & 0.0311 & -0.0983 & 0.7373
\end{bmatrix}
$$

such that $Q^{T} A = R$ is upper triangular.

Advantages and Disadvantages

The QR decomposition via Givens rotations is the most involved to implement, as the ordering of the rows required to fully exploit the algorithm is not trivial to determine. However, it has a significant advantage in that each new zero element a_{ij} affects only the row with the element to be zeroed (i) and a row above (j). This makes the Givens rotation algorithm more bandwidth efficient and parallelisable, in contrast to the Householder reflection technique.

Using Householder Reflections

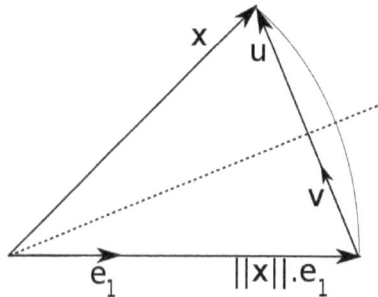

Householder reflection for QR-decomposition: The goal is to find a linear transformation that changes the vector x into a vector of same length which is collinear to e_1. We could use an orthogonal projection (Gram-Schmidt) but this will be numerically unstable if the vectors x and e_1 are close to orthogonal. Instead, the Householder reflection reflects through the dotted line (chosen to bisect the angle between x and e_1). The maximum angle with this transform is 45 degrees.

A Householder reflection (or *Householder transformation*) is a transformation that takes a vector and reflects it about some plane or hyperplane. We can use this operation to calculate the *QR* factorization of an *m*-by-*n* matrix *A* with $m \geq n$.

Q can be used to reflect a vector in such a way that all coordinates but one disappear.

Let *x* be an arbitrary real *m*-dimensional column vector of *A* such that $\| x \| = | \alpha |$ for a scalar *a*. If the algorithm is implemented using floating-point arithmetic, then *a* should get the opposite sign as the *k*-th coordinate of *x*, where x_k is to be the pivot coordinate after which all entries are 0 in matrix *A*'s final upper triangular form, to avoid loss of significance. In the complex case, set

$$\alpha = -e^{i \arg x_k} \| x \|$$

and substitute transposition by conjugate transposition in the construction of *Q* below.

Then, where e_1 is the vector $(1,0,...,0)^T$, $||\cdot||$ is the Euclidean norm and *I* is an *m*-by-*m* identity matrix, set

$$u = x - \alpha e_1,$$

$$v = \frac{u}{\| u \|},$$

$$Q = I - 2vv^T.$$

Or, if *A* is complex

$$Q = I - 2vv^H.$$

Q is an *m*-by-*m* Householder matrix and

$$Qx = (\alpha, 0, \cdots, 0)^T.$$

This can be used to gradually transform an *m*-by-*n* matrix *A* to upper triangular form. First, we multiply *A* with the Householder matrix Q_1 we obtain when we choose the first matrix column for x. This results in a matrix $Q_1 A$ with zeros in the left column (except for the first row).

$$Q_1 A = \begin{bmatrix} \alpha_1 & \star & \cdots & \star \\ 0 & & & \\ \vdots & & A' & \\ 0 & & & \end{bmatrix}$$

This can be repeated for *A'* (obtained from $Q_1 A$ by deleting the first row and first column), resulting in a Householder matrix Q'_2. Note that Q'_2 is smaller than Q_1. Since we want it really to operate on Q_1A instead of A' we need to expand it to the upper left, filling in a 1, or in general:

$$Q_k = \begin{pmatrix} I_{k-1} & 0 \\ 0 & Q'_k \end{pmatrix}.$$

After t iterations of this process, $t = \min(m-1, n)$,

$$R = Q_t \cdots Q_2 Q_1 A$$

is an upper triangular matrix. So, with

$$Q = Q_1^T Q_2^T \cdots Q_t^T,$$

$A = QR$ is a QR decomposition of A.

This method has greater numerical stability than the Gram–Schmidt method.

The following table gives the number of operations in the k-th step of the QR-decomposition by the Householder transformation, assuming a square matrix with size n.

Operation	Number of operations in the k-th step	
multiplications	$2(n-k+1)^2$	
additions	$(n-k+1)^2 + (n-k+1)(n-k)+2$	
division	1	
square root	1	

Summing these numbers over the $n-1$ steps (for a square matrix of size n), the complexity of the algorithm (in terms of floating point multiplications) is given by

$$\frac{2}{3}n^3 + n^2 + \frac{1}{3}n - 2 = O(n^3).$$

Example:

Let us calculate the decomposition of

$$A = \begin{pmatrix} 12 & -51 & 4 \\ 6 & 167 & -68 \\ -4 & 24 & -41 \end{pmatrix}.$$

First, we need to find a reflection that transforms the first column of matrix A, vector $a_1 = (12, 6, -4)^T$, into $\|a_1\| e_1 = (14, 0, 0)^T$.

Now,

$$u = x - \alpha e_1,$$

and

$$v = \frac{u}{\|u\|}.$$

Here,

$$\alpha = 14 \text{ and } x = a_1 = (12, 6, -4)^T$$

Therefore

$$u = (-2, 6, -4)^T = (2)(-1, 3, -2)^T \text{ and } v = \frac{1}{\sqrt{14}}(-1, 3, -2)^T, \text{ and then}$$

$$Q_1 = I - \frac{2}{\sqrt{14}\sqrt{14}} \begin{pmatrix} -1 \\ 3 \\ -2 \end{pmatrix} (-1 \quad 3 \quad -2)$$

$$= I - \frac{1}{7} \begin{pmatrix} 1 & -3 & 2 \\ -3 & 9 & -6 \\ 2 & -6 & 4 \end{pmatrix}$$

$$= \begin{pmatrix} 6/7 & 3/7 & -2/7 \\ 3/7 & -2/7 & 6/7 \\ -2/7 & 6/7 & 3/7 \end{pmatrix}.$$

Now observe:

$$Q_1 A = \begin{pmatrix} 14 & 21 & -14 \\ 0 & -49 & -14 \\ 0 & 168 & -77 \end{pmatrix},$$

so we already have almost a triangular matrix. We only need to zero the (3, 2) entry.

Take the (1, 1) minor, and then apply the process again to

$$A' = M_{11} = \begin{pmatrix} -49 & -14 \\ 168 & -77 \end{pmatrix}.$$

By the same method as above, we obtain the matrix of the Householder transformation

$$Q = Q_1^T Q_2^T = \begin{pmatrix} 6/7 & -69/175 & 58/175 \\ 3/7 & 158/175 & -6/175 \\ -2/7 & 6/35 & 33/35 \end{pmatrix}$$

after performing a direct sum with 1 to make sure the next step in the process works properly.

Now, we find

$$Q = Q_1^T Q_2^T = \begin{pmatrix} 0.8571 & -0.3943 & 0.3314 \\ 0.4286 & 0.9029 & -0.0343 \\ -0.2857 & 0.1714 & 0.9429 \end{pmatrix}$$

Or, to four decimal digits,

$$R = Q_2 Q_1 A = Q^T A = \begin{pmatrix} 14 & 21 & -14 \\ 0 & 175 & -70 \\ 0 & 0 & -35 \end{pmatrix}.$$

The matrix Q is orthogonal and R is upper triangular, so $A = QR$ is the required QR-decomposition.

Advantages and Disadvantages

The use of Householder transformations is inherently the most simple of the numerically stable QR decomposition algorithms due to the use of reflections as the mechanism for producing zeroes in the R matrix. However, the Householder reflection algorithm is bandwidth heavy and not parallelisable, as every reflection that produces a new zero element changes the entirety of both Q and R matrices.

Using the Gram–Schmidt Process

Consider the Gram–Schmidt process applied to the columns of the full column rank matrix $A = [a_1, \cdots, a_n]$, with inner product $\langle v, w \rangle = v^T w$ (or $\langle v, w \rangle = v^* w$ for the complex case).

Define the projection:

$$\text{proj}_u a = \frac{\langle u, a \rangle}{\langle u, u \rangle} u$$

then:

$$u_1 = a_1, \qquad\qquad e_1 = \frac{u_1}{\|u_1\|}$$

$$u_2 = a_2 - \text{proj}_{u_1} a_2, \qquad\qquad e_2 = \frac{u_2}{\|u_2\|}$$

$$u_3 = a_3 - \text{proj}_{u_1} a_3 - \text{proj}_{u_2} a_3, \quad e_3 = \frac{u_3}{\|u_3\|}$$

$$\vdots \qquad\qquad\qquad\qquad \vdots$$

$$u_k = a_k - \sum_{j=1}^{k-1} \text{proj}_{u_j} a_k, \qquad e_k = \frac{u_k}{\|u_k\|}$$

We can now express the a_i s over our newly computed orthonormal basis:

$$a_1 = \langle e_1, a_1 \rangle e_1$$
$$a_2 = \langle e_2, a_2 \rangle e_1 + \langle e_2, a_2 \rangle e_2$$
$$a_2 = \langle e_2, a_3 \rangle e_1 + \langle e_2, a_3 \rangle e_2 + \langle e_3, a_3 \rangle e_3$$
$$\vdots$$
$$a_k = \sum_{j=1}^{k} \langle e_j, a_k \rangle e_j$$

where $\langle e_i, a_i \rangle = \|u_i\|$. This can be written in matrix form:

$$A = QR$$

where:

$$Q = [e_1, \cdots, e_n]$$

and

$$R = \begin{pmatrix} \langle e_1, a_1 \rangle & \langle e_1, a_2 \rangle & \langle e_1, a_3 \rangle & \cdots \\ 0 & \langle e_2, a_2 \rangle & \langle e_2, a_3 \rangle & \cdots \\ 0 & 0 & \langle e_3, a_3 \rangle & \cdots \\ \vdots & \vdots & \vdots & \ddots \end{pmatrix}.$$

Example

Consider the decomposition of

$$A = \begin{pmatrix} 12 & -51 & 4 \\ 6 & 167 & -68 \\ -4 & 24 & -41 \end{pmatrix}.$$

Recall that an orthonormal matrix Q has the property

$$Q^T Q = I.$$

Then, we can calculate Q by means of Gram–Schmidt as follows:

$$U = \begin{pmatrix} u_1 & u_2 & u_3 \end{pmatrix} = \begin{pmatrix} 12 & -69 & -58/5 \\ 6 & 158 & 6/5 \\ -4 & 30 & -33 \end{pmatrix};$$

$$Q = \begin{pmatrix} \dfrac{u_1}{\|u_1\|} & \dfrac{u_2}{\|u_2\|} & \dfrac{u_3}{\|u_3\|} \end{pmatrix} = \begin{pmatrix} 6/7 & -69/175 & -58/175 \\ 3/7 & 158/175 & 6/175 \\ -2/7 & 6/35 & -33/35 \end{pmatrix}.$$

Thus, we have

$$Q^T A = Q^T Q R = R;$$

$$R = Q^T A = \begin{pmatrix} 14 & 21 & -14 \\ 0 & 175 & -70 \\ 0 & 0 & 35 \end{pmatrix}.$$

Relation to RQ Decomposition

The RQ decomposition transforms a matrix A into the product of an upper triangular matrix R (also known as right-triangular) and an orthogonal matrix Q. The only difference from QR decomposition is the order of these matrices.

QR decomposition is Gram–Schmidt orthogonalization of columns of A, started from the first column.

RQ decomposition is Gram–Schmidt orthogonalization of rows of A, started from the last row.

Advantages and Disadvantages

The Gram-Schmidt process is inherently numerically unstable. While the application of the projections has an appealing geometric analogy to orthogonalization, the orthogonalization itself is prone to numerical error. A significant advantage however is the ease of implementation, which makes this a useful algorithm to use for prototyping if a pre-built linear algebra library is unavailable.

Connection to a Determinant or a Product of Eigenvalues

We can use QR decomposition to find the absolute value of the determinant of a square matrix. Suppose a matrix is decomposed as $A = QR$. Then we have

$$\det(A) = \det(Q) \cdot \det(R).$$

Since Q is unitary, $|\det(Q)| = 1$. Thus,

$$|\det(A)| = |\det(R)| = \left| \prod_i r_{ii} \right|,$$

where r_{ii} are the entries on the diagonal of R.

Furthermore, because the determinant equals the product of the eigenvalues, we have

$$\left| \prod_i r_{ii} \right| = \left| \prod_i \lambda_i \right|,$$

where λ_i are eigenvalues of A.

We can extend the above properties to non-square complex matrix A by introducing the definition of QR-decomposition for non-square complex matrix and replacing eigenvalues with singular values.

Suppose a QR decomposition for a non-square matrix A:

$$A = Q \begin{pmatrix} R \\ O \end{pmatrix}, \qquad Q^* Q = I,$$

where O is a zero matrix and Q is a unitary matrix.

From the properties of SVD and determinant of matrix, we have

$$\left| \prod_i r_{ii} \right| = \prod_i \sigma_i,$$

where σ_i are singular values of A.

Note that the singular values of A and R are identical, although their complex eigenvalues may be different. However, if A is square, the following is true:

$$\prod_i \sigma_i = \left| \prod_i \lambda_i \right|.$$

In conclusion, QR decomposition can be used efficiently to calculate the product of the eigenvalues or singular values of a matrix.

Column Pivoting

QR decomposition with column pivoting introduces a permutation matrix P:

$$AP = QR \quad \Leftrightarrow \quad A = QRP^T$$

Column pivoting is useful when A is (nearly) rank deficient, or is suspected of being so. It can also improve numerical accuracy. P is usually chosen so that the diagonal elements of R are non-increasing: $|r_{11}| \geq |r_{22}| \geq \ldots \geq |r_{nn}|$. This can be used to find the (numerical) rank of A at lower computational cost than a singular value decomposition, forming the basis of so-called rank-revealing QR algorithms.

Using for Solution to Linear Inverse Problems

Compared to the direct matrix inverse, inverse solutions using QR decomposition are more numerically stable as evidenced by their reduced condition numbers.

To solve the underdetermined ($m < n$) linear problem $Ax = b$ where the matrix A has dimensions $m \times n$ and rank m, first find the QR factorization of the transpose of A: $A^T = QR$, where Q is an orthogonal matrix (i.e. $Q \quad Q \quad$), and R has a special form:

$R = \begin{bmatrix} R_1 \\ O \end{bmatrix}$. Here R_1 is a square $m \times m$ right triangular matrix, and the zero matrix has

dimension $(n-m) \times m$. After some algebra, it can be shown that a solution to the in-

verse problem can be expressed as: $x = Q \begin{bmatrix} (R_1^T)^{-1} b \\ O \end{bmatrix}$ where one may either find R_1^{-1} by

Gaussian elimination or compute $(R_1^T)^{-1} b$ directly by forward substitution. The latter technique enjoys greater numerical accuracy and lower computations.

To find a solution, $\hat{x}$, to the over determined ($m \geq n$) problem $Ax = b$ which minimizes the norm $\| A\hat{x} - b \|$, first find the QR factorization of A: $A = QR$. The solution can then be expressed as $\hat{x} = R_1^{-1}(Q_1^T b)$, where Q_1 is an $m \times n$ matrix containing the first n columns of the full orthonormal basis Q and where R_1 is as before. Equivalent to the underdetermined case, back substitution can be used to quickly and accurately find this $\hat{x}$ without explicitly inverting R_1. (Q_1 and R_1 are often provided by numerical libraries as an "economic" QR decomposition.)

Iterative Method

The term "iterative method" refers to a wide range of techniques that use successive approximations to obtain more accurate solutions to a linear system at each step. Stationary methods are older, simpler to understand and implement, but usually not as effective. Nonstationary methods are a relatively recent development; their analysis is usually harder to understand, but they can be highly effective. The nonstationary methods are based on the idea of sequences of orthogonal vectors.

Stationary iterative method: Iterative method that performs in each iteration the same operations on the current iteration vectors. Nonstationary iterative method: Iterative method that has iteration-dependent coefficients. Dense matrix: Matrix for which the number of zero elements is too small to warrant specialized algorithms. Sparse matrix: Matrix for which the number of zero elements is large enough that algorithms avoiding operations on zero elements pay off. Matrices derived from partial differential equations typically have a number of nonzero elements that is proportional to the matrix size, while the total number of matrix elements is the square of the matrix size.

The rate at which an iterative method converges depends greatly on the spectrum of the coefficient matrix. Hence, iterative methods usually involve a second matrix that transforms the coefficient matrix into one with a more favorable spectrum. The transformation matrix is called a *preconditioner*. A good preconditioner improves the convergence of the iterative method, sufficiently to overcome the extra cost of constructing and applying the preconditioner. Indeed, without a preconditioner the iterative method may even fail to converge.

Iterative Methods for Ax=B

We will solve Ax = b. Again, for simplicity, we assume that A is a regular square matrix, hence there is a unique solution.

We have to bring the equation into a fixed-point equation and we do it by splitting the matrix A = B + (A − B) with some cleverly chosen matrix B. The way to think about it is that B is the "main part" which will be chosen a "nice" matrix, and we view A as a "small perturbation" (by A − B) of this nice matrix B.

Clearly Ax = b is equivalent to,

$$Bx = (B - A)x + b$$

or

$$x = B^{-1}[(B - A)x + b] = (I - B^{-1}A)x + B^{-1}b$$

at least if B is invertible (hence we will have to choose B such that B^{-1} be simple).

The algorithm is very simple: start with an initial vector $x^{(0)}$, and iteratively generate,

$$x^{(n)} := (I - B^{-1}A)x^{(n-1)} + B^{-1}b$$

How to predict the convergence? At least is there some sufficient condition for convergence?

Let $\delta x^{(n)} := x - x^{(n)}$ be the error after the n-th iteration. From $Ax = b$ and $Bx^{(n)} = (B - A)$ $x^{(n-1)} + b$ we easily see that,

$$B(\delta x^{(n)}) = (B - A)(\delta x^{(n-1)})$$

i.e.

$$\delta x^{(n)} := (I - B^{-1}A)(\delta x^{(n-1)}) = (I - B^{-1}A)^n \, (\delta x^{(0)})$$

where $\delta x^{(0)}$ is the difference of the initial value from the solution. Hence,

$$\left\| \delta x^{(n)} \right\| \le \left\| (I - B^{-1}A)^n \right\| \left\| \delta x^{(0)} \right\|$$

It is clear that if the norm of the matrix $(I - B^{-1}A)^n$ converges to zero, then the iteration converges. Moreover, this norm gives an estimate on the speed of convergence.

Attractive Fixed Points

If an equation can be put into the form $f(x) = x$, and a solution x is an attractive fixed point of the function f, then one may begin with a point x_1 in the basin of attraction of x, and let $x_{n+1} = f(x_n)$ for $n \ge 1$, and the sequence $\{x_n\}_{n \ge 1}$ will converge to the solution x. Here x_n is the nth approximation or iteration of x and x_{n+1} is the next or $n + 1$ iteration of x. Alternately, superscripts in parentheses are often used in numerical methods, so as not to interfere with subscripts with other meanings. (For example, $x^{(n+1)} = f(x^{(n)})$.) If the function f is continuously differentiable, a sufficient condition for convergence is that the spectral radius of the derivative is strictly bounded by one in a neighborhood of the fixed point. If this condition holds at the fixed point, then a sufficiently small neighborhood (basin of attraction) must exist.

Linear Systems

In the case of a system of linear equations, the two main classes of iterative methods are the stationary iterative methods, and the more general Krylov subspace methods.

Stationary Iterative Methods

Stationary iterative methods solve a linear system with an operator approximating the original one; and based on a measurement of the error in the result (the residual), form a "correction equation" for which this process is repeated. While these methods are simple to derive, implement, and analyze, convergence is only guaranteed for a limited class of matrices.

An *iterative method* is defined by

$$x^{k+1} := \Psi(x^k), \quad k \ge 0$$

and for a given linear system $Ax = b$ with exact solution x^* the *error* by

$$e^k := x^k - x^*, \quad k \ge 0.$$

An iterative method is called *linear* if there exists a matrix $C \in \mathbb{R}^{n \times x}$ s.t.

$$e^{k+1} = Ce^k \quad \forall k \geq 0$$

and this matrix is called *iteration matrix*. An iterative method with a given iteration matrix C is called *convergent* if the following holds,

$$\lim_{k \to \infty} C^k = 0.$$

An important theorem states that for a given iterative method and its iteration matrix C it is convergent if and only if its spectral radius $\rho(C)$ is smaller than unity, i.e.,

$$\rho(C) < 1.$$

The basic iterative methods work by a splitting up the matrix A into,

$$A = M - N$$

and here the matrix M should be easily invertible. The iterative methods are now defined as,

$$Mx^{k+1} = Nx^k + b, \quad k \geq 0.$$

From this follows that the iteration matrix is given by,

$$C = I - M^{-1}A = M^{-1}N.$$

Examples:

Basic examples of stationary iterative methods use a splitting of the matrix A such as,

$$A = D - L - U, \quad D := \text{diag}((a_{ii})_i)$$

where D is only the diagonal part of A, and L is the strict lower triangular part of A. Respectively, U is the upper triangular part of A.

- Richardson method: $M := \dfrac{1}{\omega}I \quad (\omega \neq 0)$
- Jacobi method: $M := D$
- Damped Jacobi method: $M := \dfrac{1}{\omega}D \quad (\omega \neq 0)$
- Gauss–Seidel method: $M := D - L$
- Successive over-relaxation method (SOR): $M := \dfrac{1}{\omega}D - L \quad (\omega \neq 0)$
- Symmetric successive over-relaxation (SSOR):

$$M := \frac{1}{\omega(2-\omega)}(D - \omega L)D^{-1}(D - \omega U) \quad (\omega \neq \{0,2\})$$

Linear stationary iterative methods are also called relaxation methods.

Krylov Subspace Methods

Krylov subspace methods work by forming a basis of the sequence of successive matrix powers times the initial residual (the Krylov sequence). The approximations to the solution are then formed by minimizing the residual over the subspace formed. The prototypical method in this class is the conjugate gradient method (CG) which assumes that the system matrix A is symmetric positive-definite. For symmetric (and possibly indefinite) A one works with the minimal residual method (MINRES). In the case of not even symmetric matrices methods, such as the generalized minimal residual method (GMRES) and the biconjugate gradient method (BiCG), have been derived.

Convergence of Krylov Subspace Methods

Since these methods form a basis, it is evident that the method converges in N iterations, where N is the system size. However, in the presence of rounding errors this statement does not hold; moreover, in practice N can be very large, and the iterative process reaches sufficient accuracy already far earlier. The analysis of these methods is hard, depending on a complicated function of the spectrum of the operator.

Preconditioners

The approximating operator that appears in stationary iterative methods can also be incorporated in Krylov subspace methods such as GMRES (alternatively, preconditioned Krylov methods can be considered as accelerations of stationary iterative methods), where they become transformations of the original operator to a presumably better conditioned one. The construction of preconditioners is a large research area.

Gauss–Seidel Method

The Gauss-Seidel Iteration Method is an improved version of the Jacobi Iteration Method.

suppose that $x^{(0)} = \begin{bmatrix} x_1^{(0)} \\ x_2^{(0)} \\ \vdots \\ x_n^{(0)} \end{bmatrix}$ is an initial approximation to the solution x of the following system of n equations in n unknowns:

$$
\begin{array}{llllll}
E(1): & a_{11}x_1 & + & a_{12}x_2 & + & \cdots & + & a_{1n}x_n & = & b_1 \\
E(2): & a_{21}x_1 & & a_{22}x_2 & + & \cdots & + & a_{2n}x_n & = & b_2 \\
\vdots & \vdots & \vdots & \vdots & \ddots & \ddots & & \vdots & & \vdots \\
E(n): & a_{n1}x_1 & + & a_{n2}x_2 & + & \cdots & + & a_{nn}x_n & = & b_n
\end{array}
$$

For the Gauss-Seidel Method, we once again isolate the variable x_i from equation $E(i)$ for $i = 1, 2, \ldots, n$.

$$x_1 = \frac{b_1 - [a_{12}x_2 + a_{13}x_3 + \ldots + a_{1n}x_n]}{a_{11}}$$

$$x_2 = \frac{b_2 - [a_{21}x_1 + a_{23}x_3 + \ldots + a_{2n}x_n]}{a_{22}}$$

$$\vdots$$

$$x_n = \frac{b_n - [a_{n1}x_1 + a_{n2}x_2 + \ldots + a_{n,n-1}x_{n-1}]}{a_{nn}}$$

We will now obtain a first approximation to the solution x of the actual solution x by using the Gauss-Seidel Iteration Method. We compute $x_1^{(1)}$ by plugging in the values of our initial solution approximation $x^{(0)}$. We then obtain an approximation to the entry of $x_1^{(1)}$. We use this entry and the remaining entries from $x^{(0)}$ to obtain an approximation of the entry $x_2^{(1)}$. We then use both $x_1^{(1)}$ and $x_2^{(1)}$ as well as the remaining entries from $x^{(0)}$ to obtain an approximation of the entry $x_3^{(1)}$ and so forth, and thus:

$$x_1^{(1)} = \frac{b_1 - [a_{12}x_2^{(0)} + a_{13}x_3^{(0)} + a_{14}x_4^{(0)} + \ldots + a_{1n}x_n^{(0)}]}{a_{11}}$$

$$x_2^{(1)} = \frac{b_2 - [a_{21}x_1^{(1)} + a_{23}x_3^{(0)} + a_{24}x_4^{(0)} + \ldots + a_{2n}x_n^{(0)}]}{a_{22}}$$

$$x_3^{(1)} = \frac{b_3 - [a_{31}x_1^{(1)} + a_{32}x_2^{(1)} + a_{34}x_4^{(0)} + \ldots + a_{n,n-1}x_{n-1}^{(1)}]}{a_{33}}$$

$$\vdots$$

$$x_n^{(1)} = \frac{b_n - [a_{n1}x_1^{(1)} + a_{n2}x_2^{(1)} + a_{n3}x_3^{(1)} + \ldots + a_{n,n-1}x_{n-1}]}{a_{nn}}$$

In sigma notation we have that each component $^{(1)}$ for $i = 1, 2, \ldots, n$ of $x^{(1)}$ is given by:

$$x_i^{(1)} = \frac{b_i - \left[\sum_{j=1}^{i-1} a_{ij}x_j^{(1)} + \sum_{j=i+1}^{n} a_{ij}x_j^{(0)} \right]}{a_{ii}}$$

To obtain the second approximation $x^{(2)}$ of x using the Gauss-Seidel method, we would have that:

$$x_1^{(2)} = \frac{b_1 - \left[a_{12}x_2^{(1)} + a_{13}x_3^{(1)} + a_{14}x_4^{(1)} + \ldots + a_{1n}x_n^{(1)} \right]}{a_{11}}$$

$$x_2^{(2)} = \frac{b_2 - \left[a_{12}x_1^{(2)} + a_{23}x_3^{(1)} + a_{24}x_4^{(1)} + \ldots + a_{2n}x_n^{(1)} \right]}{a_{22}}$$

$$x_3^{(2)} = \frac{b_n - \left[a_{31}x_1^{(2)} + a_{32}x_2^{(2)} + a_{34}x_4^{(1)} + \ldots + a_{3n}x_n^{(1)} \right]}{a_{33}}$$

$$\vdots$$

$$x_n^{(2)} = \frac{b_n - \left[a_{n1}x_1^{(2)} + a_{n2}x_2^{(2)} + a_{n3}x_3^{(2)} + \ldots + a_{n,n-1}x_{n-1}^{(2)} \right]}{a_{nn}}$$

In sigma notation we have that each component $x_i^{(2)}$ for $i = 1, 2, \ldots, n$ of $x^{(2)}$ is given by:

$$x_i^{(2)} = \frac{b_i - \left[\sum_{j=1}^{i-1} a_{ij}x_j^{(2)} + \sum_{j=i+1}^{n} a_{ij}x_j^{(1)} \right]}{a_{ii}}$$

We can continue approximating x with these solutions in the hopes that the sequence of approximations with the Gauss-Seidel method converges to the true solution. Thus for k≥1 and for $i = 1, 2, \ldots, n$, the kth iteration of the Gauss-Seidel method yields:

$$x_1^{(k)} = \frac{b_1 - \left[a_{12}x_2^{(k-1)} + a_{13}x_3^{(k-1)} + a_{14}x_4^{(k-1)} + \ldots + a_{1n}x_n^{(k-1)} \right]}{a_{11}}$$

$$x_2^{(k)} = \frac{b_2 - \left[a_{21}x_1^{(k)} + a_{23}x_3^{(k-1)} + a_{24}x_4^{(k-1)} + \ldots + a_{2n}x_n^{(k-1)} \right]}{a_{11}}$$

$$x_3^{(k)} = \frac{b_3 - \left[a_{31}x_1^{(k)} + a_{32}x_2^{(k-1)} + a_{34}x_4^{(k-1)} + \ldots + a_{3n}x_n^{(k-1)} \right]}{a_{11}}$$

$$\vdots$$

$$x_n^{(k)} = \frac{b_n - \left[a_{n1}x_1^{(k)} + a_{n2}x_2^{(k)} + a_{n3}x_3^{(k)} + \ldots + a_{n,n-1}x_{n-1}^{(k)} \right]}{a_{11}}$$

In sigma notation we have that each entry $x_i^{(k)}$ for $i = 1, 2, \ldots, n$ in the approximation $x^{(k)}$ of x is given by:

$$x_i^{(k)} = \frac{b_i - \left[\sum_{j=1}^{i-1} a_{ij}x_j^{(k)} + \sum_{j=i+1}^{n} a_{ij}x_j^{(k-1)} \right]}{a_{ii}}$$

Numerical Algorithm of Gauss-seidel Method

Input: $A = [a_{ij}], b, XO = x^{(o)}$, tolerance *TOL*, maximum number of iterations *N*.

Step 1: Set $k = 1$

Step 2: while $(k \le N)$ do Steps 3-6

Step 3 For for $i = 1, 2,, \ldots n$

$$x_i = \frac{1}{a_{ii}} [-\sum_{j=1}^{i-1} (a_{ij} x_j) - \sum_{j=i+1}^{n} (a_{ij} x O_j) + b_i],$$

Step 4: If $\| x - XO \| < TOL,$ then OUTPUT $(x_1, x_2, x_3, \ldots x_n)$; STOP.

Step 5: Set $k = k + 1.$

Step 6: For for $i = 1, 2,, \ldots n$

Set $XO_i = x_i.$

Step 7: OUTPUT $(x_1, x_2, x_3, \ldots x_n)$;

STOP.

Convergence Theorems of the Iteration Methods

Let the iteration method be written as $x^{(k)} = Tx^{(k-1)} + c$ for each $k = 1, 2, \ldots$

Lemma: If the spectral radius satisfies $\rho(T) < 1,$ then $(I - T)^{-1}$ exists, and

$$(I - T)^{-1} = I + T + T^2 + \ldots = \sum_{j=0}^{\infty} T^j$$

Theorem: For any $x^{(0)} \in R^n,$ the sequence $\{x^{(k)}\}_{k=1}^{\infty}$ defined by,

$$x^{(k)} = Tx^{(k-1)} + c \quad \text{for each } k \geq 1$$

converges to the unique solution of $x = Tx + c$ if and only if $\rho(T) < 1.$

Proof (only show $\rho(T) < 1$ is sufficient condition)

$$x^{(k)} = Tx^{(k-1)} + c = T(Tx^{(k-2)} + c) + c = \cdots = T^k x^{(0)} + (T^{(k-1)} + \cdots + T + I)c$$

Since $\rho(T) < 1, \lim_{k \to \infty} T^k x^{(0)} = 0$

$$\lim_{k \to \infty} x^{(k)} = 0 + \lim_{k \to \infty} \left(\sum_{j=0}^{k-1} T^j \right) c \; (I - T)^{-1} c$$

Corollary: If $\| T \| < 1$ for any natural matrix norm and c is a given vector, then the sequence $\{x^{(k)}\}_{k=0}^{\infty}$ defined by $x^{(k)} = Tx^{(k-1)} + c$ converges, for any $X^{(0)} \in R^n,$ to a vector $X \in R^n,$ with $x = Tx + c,$ and the following error bound hold:

1. $\| x - x^{(k)} \| \leq \| T \|^k \| x^{(0)} - x \|$

2. $\left\| x - x^{(k)} \right\| \leq \dfrac{\left\| T \right\|^k}{1 - \left\| T \right\|} \left\| x^{(1)} - x^{(0)} \right\|$

Theorem: If A is strictly diagonally dominant, then for any choice of $x^{(0)}$, both the Jacobi and Gauss-Seidel methods give sequences $\{x^{(k)}\}_{k=0}^{\infty}$ that converges to the unique solution of $Ax = b$.

Rate of Convergence

Corollary: (i) implies $\left\| x - x^{(k)} \right\| \approx \rho(T)^k \left\| x^{(0)} - x \right\|$

1. Theorem (Stein-Rosenberg) If $a_{ij} \leq 0$, for each $i \neq j$ and $a_{ii} \geq 0$, for each $i = 1, 2, \ldots, n$, then one and only one of following statements holds:

2. $0 \leq \rho(T_g) < \rho(T_j) < 1$;

3. $1 < \rho(T_j) < \rho(T_g)$;

4. $\rho(T_j) = \rho(T_g) = 0$;

5. $\rho(T_j) = \rho(T_g) = 1$.

Example for the Matrix Version

A linear system shown as $Ax = b$ is given by:

$$A = \begin{bmatrix} 16 & 3 \\ 7 & -11 \end{bmatrix} \text{ and } b = \begin{bmatrix} 11 \\ 13 \end{bmatrix}.$$

We want to use the equation,

$$x^{(k+1)} = L_*^{-1}(b - Ux^{(k)})$$

in the form,

$$x^{(k+1)} = Tx^{(k)} + C$$

where:

$$T = -L_*^{-1}U \text{ and } C = L_*^{-1}b.$$

We must decompose A into the sum of a lower triangular component L_* and a strict upper triangular component U:

$$L_* = \begin{bmatrix} 16 & 0 \\ 7 & -11 \end{bmatrix} \text{ and } U = \begin{bmatrix} 0 & 3 \\ 0 & 0 \end{bmatrix}.$$

The inverse of L_* is:

$$L_*^{-1} = \begin{bmatrix} 16 & 0 \\ 7 & -11 \end{bmatrix}^{-1} = \begin{bmatrix} 0.0625 & 0.0000 \\ 0.0398 & -0.0909 \end{bmatrix}.$$

Now we can find:

$$T = -\begin{bmatrix} 0.0625 & 0.0000 \\ 0.0398 & -0.0909 \end{bmatrix} \times \begin{bmatrix} 0 & 3 \\ 0 & 0 \end{bmatrix} = \begin{bmatrix} 0.000 & -0.1875 \\ 0.000 & -0.1193 \end{bmatrix},$$

$$C = \begin{bmatrix} 0.0625 & 0.0000 \\ 0.0398 & -0.0909 \end{bmatrix} \times \begin{bmatrix} 11 \\ 13 \end{bmatrix} = \begin{bmatrix} 0.6875 \\ -0.7443 \end{bmatrix}.$$

Now we have T and C and we can use them to obtain the vectors x iteratively.

First of all, we have to choose $x^{(0)}$: we can only guess. The better the guess, the quicker the algorithm will perform.

We suppose:

$$x^{(0)} = \begin{bmatrix} 1.0 \\ 1.0 \end{bmatrix}.$$

We can then calculate:

$$x^{(1)} = \begin{bmatrix} 0.000 & -0.1875 \\ 0.000 & -0.1193 \end{bmatrix} \times \begin{bmatrix} 1.0 \\ 1.0 \end{bmatrix} + \begin{bmatrix} 0.6875 \\ -0.7443 \end{bmatrix} = \begin{bmatrix} 0.5000 \\ -0.8636 \end{bmatrix}.$$

$$x^{(2)} = \begin{bmatrix} 0.000 & -0.1875 \\ 0.000 & -0.1193 \end{bmatrix} \times \begin{bmatrix} 0.5000 \\ -0.8636 \end{bmatrix} + \begin{bmatrix} 0.6875 \\ -0.7443 \end{bmatrix} = \begin{bmatrix} 0.8494 \\ -0.6413 \end{bmatrix}.$$

$$x^{(3)} = \begin{bmatrix} 0.000 & -0.1875 \\ 0.000 & -0.1193 \end{bmatrix} \times \begin{bmatrix} 0.8494 \\ -0.6413 \end{bmatrix} + \begin{bmatrix} 0.6875 \\ -0.7443 \end{bmatrix} = \begin{bmatrix} 0.8077 \\ -0.6678 \end{bmatrix}.$$

$$x^{(4)} = \begin{bmatrix} 0.000 & -0.1875 \\ 0.000 & -0.1193 \end{bmatrix} \times \begin{bmatrix} 0.8077 \\ -0.6678 \end{bmatrix} + \begin{bmatrix} 0.6875 \\ -0.7443 \end{bmatrix} = \begin{bmatrix} 0.8127 \\ -0.6646 \end{bmatrix}.$$

$$x^{(5)} = \begin{bmatrix} 0.000 & -0.1875 \\ 0.000 & -0.1193 \end{bmatrix} \times \begin{bmatrix} 0.8127 \\ -0.6646 \end{bmatrix} + \begin{bmatrix} 0.6875 \\ -0.7443 \end{bmatrix} = \begin{bmatrix} 0.8121 \\ -0.6650 \end{bmatrix}.$$

$$x^{(6)} = \begin{bmatrix} 0.000 & -0.1875 \\ 0.000 & -0.1193 \end{bmatrix} \times \begin{bmatrix} 0.8121 \\ -0.6650 \end{bmatrix} + \begin{bmatrix} 0.6875 \\ -0.7443 \end{bmatrix} = \begin{bmatrix} 0.8122 \\ -0.6650 \end{bmatrix}.$$

$$x^{(7)} = \begin{bmatrix} 0.000 & -0.1875 \\ 0.000 & -0.1193 \end{bmatrix} \times \begin{bmatrix} 0.8122 \\ -0.6650 \end{bmatrix} + \begin{bmatrix} 0.6875 \\ -0.7443 \end{bmatrix} = \begin{bmatrix} 0.8122 \\ -0.6650 \end{bmatrix}.$$

As expected, the algorithm converges to the exact solution:

$$x = A^{-1}b \approx \begin{bmatrix} 0.8122 \\ -0.6650 \end{bmatrix}.$$

In fact, the matrix A is strictly diagonally dominant (but not positive definite).

Another Example for the Matrix Version

Another linear system shown as $Ax = b$ is given by:

$$A = \begin{bmatrix} 2 & 3 \\ 5 & 7 \end{bmatrix} \text{ and } b = \begin{bmatrix} 11 \\ 13 \end{bmatrix}.$$

We want to use the equation,

$$x^{(k+1)} = L_*^{-1}(b - Ux^{(k)})$$

in the form,

$$x^{(k+1)} = Tx^{(k)} + C$$

where:

$$T = -L_*^{-1}U \text{ and } C = L_*^{-1}b.$$

We must decompose A into the sum of a lower triangular component L_* and a strict upper triangular component U:

$$L_* = \begin{bmatrix} 2 & 0 \\ 5 & 7 \end{bmatrix} \text{ and } U = \begin{bmatrix} 0 & 3 \\ 0 & 0 \end{bmatrix}.$$

The inverse of L_* is:

$$L_*^{-1} = \begin{bmatrix} 2 & 0 \\ 5 & 7 \end{bmatrix}^{-1} = \begin{bmatrix} 0.500 & 0.000 \\ -0.357 & 0.143 \end{bmatrix}.$$

Now we can find:

$$T = -\begin{bmatrix} 0.500 & 0.000 \\ -0.357 & 0.143 \end{bmatrix} \times \begin{bmatrix} 0 & 3 \\ 0 & 0 \end{bmatrix} = \begin{bmatrix} 0.000 & -1.500 \\ 0.000 & 1.071 \end{bmatrix},$$

$$C = \begin{bmatrix} 0.500 & 0.000 \\ -0.357 & 0.143 \end{bmatrix} \times \begin{bmatrix} 11 \\ 13 \end{bmatrix} = \begin{bmatrix} 5.500 \\ -2.071 \end{bmatrix}.$$

Now we have T and C and we can use them to obtain the vectors x iteratively.

First of all, we have to choose $x^{(0)}$: we can only guess. The better the guess, the quicker will perform the algorithm.

We suppose:

$$x^{(0)} = \begin{bmatrix} 1.1 \\ 2.3 \end{bmatrix}.$$

We can then calculate:

$$x^{(1)} = \begin{bmatrix} 0 & -1.500 \\ 0 & 1.071 \end{bmatrix} \times \begin{bmatrix} 1.1 \\ 2.3 \end{bmatrix} + \begin{bmatrix} 5.500 \\ -2.071 \end{bmatrix} = \begin{bmatrix} 2.050 \\ 0.393 \end{bmatrix}.$$

$$x^{(2)} = \begin{bmatrix} 0 & -1.500 \\ 0 & 1.071 \end{bmatrix} \times \begin{bmatrix} 2.050 \\ 0.393 \end{bmatrix} + \begin{bmatrix} 5.500 \\ -2.071 \end{bmatrix} = \begin{bmatrix} 4.911 \\ -1.651 \end{bmatrix}.$$

$$x^{(3)} = \cdots.$$

If we test for convergence we'll find that the algorithm diverges. In fact, the matrix A is neither diagonally dominant nor positive definite. Then, convergence to the exact solution,

$$x = A^{-1}b = \begin{bmatrix} -38 \\ 29 \end{bmatrix}$$

is not guaranteed and, in this case, will not occur.

An Example for the Equation Version

Suppose given k equations where x_n are vectors of these equations and starting point x_0. From the first equation solve for x_1 in terms of $x_{n+1}, x_{n+2}, \ldots, x_n$. For the next equations substitute the previous values of xs.

To make it clear let's consider an example,

$$\begin{aligned} 10x_1 & -x_2 & +2x_3 & & = 6, \\ -x_1 & +11x_2 & -x_3 & +3x_4 & = 25, \\ 2x_1 & -x_2 & +10x_3 & -x_4 & = -11, \\ & 3x_2 & -x_3 & +8x_4 & = 15. \end{aligned}$$

Solving for x_1, x_2, x_3 and x_4 gives:

$$x_1 = x_2/10 - x_3/5 + 3/5,$$
$$x_2 = x_1/11 + x_3/11 - 3x_4/11 + 25/11,$$
$$x_3 = -x_1/5 + x_2/10 + x_4/10 - 11/10,$$
$$x_4 = -3x_2/8 + x_3/8 + 15/8.$$

Suppose we choose (0, 0, 0, 0) as the initial approximation, then the first approximate solution is given by,

$$x_1 = 3/5 = 0.6,$$
$$x_2 = (3/5)/11 + 25/11 = 3/55 + 25/11 = 2.3272,$$
$$x_3 = -(3/5)/5 + (2.3272)/10 - 11/10 = -3/25 + 0.23272 - 1.1 = -0.9873,$$
$$x_4 = -3(2.3272)/8 + (-0.9873)/8 + 15/8 = 0.8789.$$

Using the approximations obtained, the iterative procedure is repeated until the desired accuracy has been reached. The following are the approximated solutions after four iterations,

x_1	x_2	x_3	x_4
0.6	2.32727	-0.987273	0.878864
1.03018	2.03694	-1.01446	0.984341
1.00659	2.00356	-1.00253	0.998351
1.00086	2.0003	-1.00031	0.99985

The exact solution of the system is (1, 2, -1, 1).

Inverse Iteration

Inverse iteration is a well-known iterative procedure to compute approximations of eigenfunctions and eigenvalues of linear operators. To treat the matrix eigenvalue problem,

$$Ax_i = \lambda_i x_i$$

for a real or complex square matrix A. The scalar λ_i is the ith eigenvalue and the vector xi denotes a corresponding eigenvector. Given a nonzero starting vector $x^{(0)}$, inverse iteration generates a sequence of iterates $x^{(k)}$ by solving the linear systems,

$$(A - \sigma I)x^{(k+1)} = x^{(k)}, \quad k = 0, 1, 2, \ldots,$$

where σ denotes an eigenvalue approximation and I is the identity matrix. In practice, the iterates are normalized after each step. If A is a symmetric matrix, then the iterates $x^{(k)}$ converge to an eigenvector associated with an eigenvalue closest to σ if the starting vector $x^{(0)}$ is not perpendicular to that vector.

Convergence of inverse iteration toward an eigenvector can be estimated in terms of the Rayleigh quotients of the iterates. The Rayleigh quotient of a vector x is given by

$$\lambda(x) = \frac{(x, Ax)}{(x, x)},$$

where $(\cdot, \cdot)$ denotes the Euclidean product. The eigenvectors are the stationary points of $\lambda(\cdot)$ and its absolute extrema are the extremal eigenvalues of A.

Convergence Estimates for Inverse Iteration

To derive sharp convergence estimates for the Rayleigh quotient in the case of inverse iteration being restricted to a symmetric positive definite matrix A. This restrictive assumption is typically fulfilled for an important class of (extremely) large eigenproblems, i.e., discretizations of certain elliptic partial differential operators. These convergence estimates show that inverse iteration for a symmetric positive definite matrix and under a certain assumption on the shift parameter σ is a descent scheme for the Rayleigh quotient.

Let us first make the point that usually, the convergence theory of inverse iteration is founded on an eigenvector expansion of the initial vector, i.e., applying $(A - \sigma I)^{-1}$ to the actual iteration vector results in a relative amplification of the eigenvector components corresponding to eigenvalues close to σ . Such a convergence analysis does not exploit any properties of the Rayleigh quotient. But there is a different way to look at inverse iteration which is initiated by the demand that today, one is faced with the problem to solve extremely large eigenproblems in which the dimension of A exceeds, say, 106 up to 109 . Such matrix eigenproblems appear for instance as mesh discretizations of self-adjoint, elliptic partial differential operators. Typically, only a few of the smallest eigenvalues together with the eigenvalues are to be computed. Inverse iteration can be applied.

For these approximate versions of inverse iteration (called "inexact inverse iteration" or "preconditioned inverse iteration") any convergence theory built on an eigenvector expansion of the initial vector breaks down because an approximate solution of (1) may weed out certain eigenvector components and may amplify others in a complex and hardly controllable way. Nevertheless, as it turned out in the convergence analysis of these methods, the Rayleigh quotient can serve as a robust convergence measure since one can prove its stepwise monotonous decrease.

The technique of proof is rather unusual: the Lagrange multiplier method is applied to determine constrained extrema of the Rayleigh quotient with respect to its level sets. By doing so we first obtain the justification to restrict the analysis to two-dimensional A-invariant subspaces. In a second step we derive the convergence estimates by means of a mini-dimensional (2D) analysis. The convergence of the Rayleigh quotients is measured in terms of the ratios,

$$\Delta_{i,i+1}(\lambda) := \frac{\lambda - \lambda_i}{\lambda_{i+1} - \lambda} \text{ and } \Delta_{1,n}(\lambda) := \frac{\lambda - \lambda_1}{\lambda_n - \lambda}$$

$\lambda_2 < \ldots < \lambda_n$. A small ratio $\Delta_{i,i+1}(\lambda)$, for instance $0 \le \Delta_{i,i+1}(\lambda) \le \epsilon$ with $\epsilon > 0$, is an absolute measure for the closeness of λ to the eigenvalue λ_i, as then $\lambda \le (\lambda_1 + \epsilon \lambda_{i+1})/(1 + \epsilon) = l_1 + \mathcal{O}(\epsilon)$.

Theorem: Let $A \in \mathbb{R}^{s \times s}$ be a symmetric positive definite matrix with the eigenvalues $0 < \lambda_1 < \ldots < \lambda_n$; the multiplicity of λ_i is denoted by mi so that $m_1 + \ldots + m_n = S$.

For any real number $\lambda \in (\lambda_i, \lambda_{i+1})$ let

$$L(\lambda) = \{x \in \mathbb{R}^s ; \lambda(x) = \lambda\}$$

which is a level set of the Rayleigh quotient. Moreover, assume also the shifted matrix $A - \sigma I$ positive definite, i.e. $\sigma \in [0, \lambda_1]$,

Then for any $x \in L(\lambda)$ the next iterate $x' = (A - \sigma I)^{-1}x$ of (1) with the Rayleigh quotient $\lambda(x') = \lambda\left((A - \sigma I)^{-1}x\right)$ satisfies,

$$\Delta_{i,i+1}(\lambda(x')) \le (\rho_{i,i+1})^2 \Delta_{i,i+1}(\lambda).$$

Inequality above is an estimate on the poorest convergence of $\lambda(x')$ toward the closest eigenvalue $\lambda_i < \lambda$ in terms of ($\Delta_{1,n}(\lambda) := \frac{\lambda - \lambda_1}{\lambda_n - \lambda}$) and the convergence facto $\rho_{i,i+1}$ which is defined by (4). The right-hand side of above equation does not depend on the choice of x, but only on λ. is sharp as it is attained in a certain $x \in L(\lambda)$.

Moreover, the fastest convergence is described by a sharp estimate from below

$$(\rho_{1,n})^2 \Delta_{i,n}(\lambda) \le \Delta_{1,n}(\lambda(x')).$$

Once again, there is certain $x \in L(\lambda)$ in which the lower bound $((\rho_{1,n})^2 \Delta_{i,n}(\lambda) \le \Delta_{1,n}(\lambda(x'))$. is attained.

For $\sigma = 0$ the following sharp estimate for $\lambda(x')$ results from $\Delta_{i,i+1}(\lambda(x')) \le (\rho_{i,i+1})^2 \Delta_{i,i+1}(\lambda)$. and $(\rho_{1,n})^2 \Delta_{i,n}(\lambda) \le \Delta_{1,n}(\lambda(x'))$.

$$\frac{1}{\lambda_1^{-1} + \lambda_n^{-1} - (\lambda_1 + \lambda_n - \lambda)^{-1}} \le \lambda(x') \le \frac{1}{\lambda_i^{-1} + \lambda_{i+1}^{-1} - (\lambda_1 + \lambda_{i+1} - \lambda)^{-1}}.$$

Remark: If the initial iterate $x^{(0)}$ satisfies $\lambda(x^{(0)}) < \lambda_2$, then $(\Delta_{i,i+1}(\lambda(x')) \leq (\rho_{i,i+1})^2 \Delta_{i,i+1}(\lambda))$. can be applied,

$$\frac{\Delta_{1,2}(\lambda(x^{(k)}))}{\Delta_{1,2}(\lambda(x^{(0)}))} \leq \left(\frac{\lambda_1 - \sigma}{\lambda_2 - \sigma}\right)^{2k}, \quad k = 1, 2, \ldots,$$

and guarantees convergence of $(x^{(k)} / \|x^{(k)}\|, \lambda(x^{(k)}))$ to an eigenpair (x_1, λ_1).

Note that theorem above does not refer to the components of an eigenvector expansion of the initial vector $x^{(0)}$. Consequently, as reflected by the estimate $\Delta_{i,i+1}(\lambda(x')) \leq (\rho_{i,i+1})^2 \Delta_{i,i+1}(\lambda)$, inverse iteration starting with $\lambda(x^{(0)}) \in (\lambda_i, \lambda_{i+1})$ in the case of poorest convergence can only be shown to converge to an eigenpair (x_i, λ_i).

Proof: Let

$$U^T A U = \text{diag}\,(\underbrace{\lambda_1, \ldots, \lambda_1}_{m_1}, \underbrace{\lambda_2, \ldots, \lambda_2}_{m_2}, \ldots, \underbrace{\lambda_n, \ldots, \lambda_n}_{m_n},) =: \wedge \in \mathbb{R}^{s \times s}$$

be a diagonal matrix with λ_i being the ith eigenvalue of A with the multiplicity m_i. Then for any $x \in L(\lambda)$ one obtains $v = U^T x$ as the corresponding coefficient vector with respect to the eigenbasis. It holds $\lambda(x) = (v, \wedge v)/(v, v) =: \lambda(v)$ and

$$\lambda(x') = \lambda\,((A - \sigma I)^{-1} x) = \frac{(v, \wedge (\wedge - \sigma I)^{-2} v)}{(v, (\wedge - \sigma I)^{-2} v)} =: \lambda((\wedge - \sigma I)^{-1} v),$$

where we use the same notation $\lambda(\cdot)$ for the Rayleigh quotient with respect to both bases.

Next we give a justification for restricting the analysis to simple eigenvalues only. Therefore, let $\bar{\wedge} := \text{diag}\,(\lambda_1, \ldots, \lambda_n) \in \mathbb{R}^{n \times n}$. For any $v \in \mathbb{R}^s$ with $\lambda = \lambda(v)$ define $\bar{v} \in \mathbb{R}^n$ in such a way that,

$$\bar{v}_i = \left(\sum_{l=m+1}^{m+m_i} v_l^2\right)^{1/2} \text{ with } m = m_1 + m_2 + \ldots + m_{i-1}, \quad i = 1, \ldots, n, \; m_0 = 0,$$

i.e., all components of v corresponding to λ_1 are condensed into the single component $\bar{v}_i$. Then,

$$\bar{\lambda}(\bar{v}) := \frac{(\bar{v}, \bar{\wedge}\,\bar{v})}{(\bar{v}, \bar{v})} = \frac{(v, \wedge v)}{(v, v)} = \lambda$$

and

$$\bar{\lambda}((\bar{\wedge} - \sigma I_{n \times n})^{-1} \bar{v}) := \frac{(\bar{v}, \bar{\wedge}(\bar{\wedge} - \sigma I_{n \times n})^{-2} \bar{v})}{(\bar{v}, (\bar{\wedge} - \sigma I_{n \times n})^{-2} \bar{v})} = \frac{(v, \wedge (\wedge - \sigma I_{s \times s})^{-2} v)}{(v, (\wedge - \sigma I_{s \times s})^{-2} v)} = \lambda(x'),$$

which is a representation of the Rayleigh quotient,

$$\lambda(x') = \lambda\left((A - \sigma I)^{-1}x\right) = \frac{(v, \wedge(\wedge - \sigma I)^{-2}v)}{(v, (\wedge - \sigma I)^{-2}v} =: \lambda((\wedge - \sigma I)^{-1}v), \text{ in terms of the reduced matrix}$$

$\overline{\wedge}$ with only simple eigenvalues. This justifies to assume $m_i = 1, i = 1, \ldots, n$ in the following. Thus $s = n$ and $\overline{\wedge} = \wedge$.

A necessary condition for $\lambda(x') = \lambda\left((A - \sigma I)^{-1}x\right) = \frac{(v, \wedge(\wedge - \sigma I)^{-2}v)}{(v, (\wedge - \sigma I)^{-2}v} =: \lambda((\wedge - \sigma I)^{-1}v),$

being an extremum on the level set $L(\lambda)$ can be derived by means of the Lagrange multiplier method. Let us reformulate the non-quadratic constraint $\lambda(v) = \lambda$ as a quadratic normalization condition, i.e. $(v, v) = 1$, and the quadratic constraint $(v, \wedge v) = \lambda$. Then we consider the Lagrange function,

$$\mathcal{L}(v) = \frac{(v, \wedge(\wedge - \sigma I)^{-2}v)}{(v, (\wedge - \sigma I)^{-2}v)} + \mu((v, v) - 1) + \upsilon((v, \wedge v) - \lambda),$$

with μ and υ being the Lagrange multipliers. Any constrained extremum in v has to satisfy the equation,

$$\nabla\mathcal{L}(v) = \frac{2}{(v, (\wedge - \sigma I)^{-2}v)}(\wedge - \sigma I)^{-2}[\wedge - \lambda'I]v + 2\mu v + 2\upsilon \wedge v = 0$$

With $\lambda' := \lambda\left((\wedge - \sigma I)^{-1}v\right)$. Since v is not an eigenvector (as $\lambda \neq \lambda_i, i = 1, \ldots, n$), there are at least two nonzero components v_k and v_l with $k \neq l$. Take k as the smallest index with $v_k \neq 0$ and l as the largest index with $v_k \neq 0$. Then $\lambda_k < \lambda'$. We determine the Lagrange multipliers μ and υ from equation $(\nabla\mathcal{L}(v) = \frac{2}{(v, (\wedge - \sigma I)^{-2}v)}(\wedge - \sigma I)^{-2}[\wedge - \lambda'I]v + 2\mu v + 2\upsilon \wedge v = 0)$ by solving the linear system,

$$\begin{pmatrix} 1 & \lambda_k \\ 1 & \lambda_l \end{pmatrix}\begin{pmatrix} \mu \\ \upsilon \end{pmatrix} = \frac{1}{(v, (\wedge - \sigma I)^{-2}v)}\begin{pmatrix} (\lambda' - \lambda_k)(\lambda_k - \sigma)^{-2} \\ (\lambda' - \lambda_l)(\lambda_l - \sigma)^{-2} \end{pmatrix}$$

having a non-vanishing determinant. Its solution reads,

$$\mu = [\sigma^2\lambda' + 2\sigma(\lambda_k\lambda_l - \lambda_k\lambda' - \lambda_l\lambda') + \lambda_l^2(\lambda' - \lambda_k) + \lambda_k^2(\lambda' - \lambda_l) + \lambda_k\lambda_l\lambda']/C,$$

$$\upsilon = -[\sigma^2 - 2\sigma\lambda' + \lambda'(\lambda_k + \lambda_l) - \lambda_k\lambda_l]/C,$$

with $C = (v, (\wedge - \sigma I)^{-2}v)(\lambda_k - \sigma)^2(\lambda_l - \sigma)^2$. To show that υ has exactly two nonzero components, i.e., $v_j = 0$ for $j \neq k, l$, we insert μ and υ in the jth component of $(\nabla\mathcal{L}(v) = \frac{2}{(v, (\wedge - \sigma I)^{-2}v)}(\wedge - \sigma I)^{-2}[\wedge - \lambda'I]v + 2\mu v + 2\upsilon \wedge v = 0)$. We write $(\nabla\mathcal{L}(v))_j = \alpha(\sigma)p(\sigma)v_j,$

where $\alpha(\sigma) = (\lambda_l - \lambda_j)(\lambda_k + \lambda_j)/(C(\lambda_j - \sigma)^2 \neq 0$ and

$$p(\sigma) = 2\sigma^3 - \sigma^2\,(\lambda_k + \lambda_l + \lambda_j + 3\lambda') + 2\sigma\lambda'\,(\lambda_k + \lambda_l + \lambda_j) + \lambda_j\lambda_k\lambda_l - \lambda'\,(\lambda_k\lambda_l + \lambda_k\lambda_j + \lambda_l\lambda_j).$$

It remains to be shown that $p(\sigma) \neq 0$, where by assumption $\sigma \in (0, \lambda_1)$. First notice that $0 \leq \sigma < \lambda_1 \leq \lambda_k < \lambda' < \lambda_l$ and $\lambda_k < \lambda_j < \lambda_l$ as well as $\lim_{\sigma \to \infty} p(\sigma) = -\infty$. Moreover, the local extrema of ( ), i.e., $p'(\sigma) = 0$, are taken in λ' and $(\lambda_k + \lambda_l + \lambda_j)/3$ and are both larger than λ_k. Finally, we conclude with,

$$p(\lambda_k) = -(\lambda_j - \lambda_k)\,(\lambda_l - \lambda_k)(\lambda' - \lambda_k) < 0,$$

that it is impossible for the third order polynomial $p(\sigma)$ to take a zero in $[0, \lambda_l)$.

Thus the further ("mini-dimensional"-) analysis can be restricted to the 2D space spanned by the eigenvectors to λ_k and λ_l. The nonzero components v_k and v_l are determined by $(v, v) = 1$ and $(v, \Lambda v) = \lambda$. We obtain,

$$v_k^2 = \frac{\lambda_l - \lambda}{\lambda_l - \lambda_k}, \quad \text{and} \quad v_l^2 = \frac{\lambda - \lambda_k}{\lambda_l - \lambda_k}.$$

Inserting above equation in $\lambda' = \lambda\,((\wedge - \sigma I)^{-1}v)$ results in,

$$\lambda' = \lambda'\,(\lambda_k, \lambda_l, \lambda, \sigma) = \frac{\sigma^2\lambda - 2\sigma\lambda_k\lambda_l + \lambda_k\lambda_l(\lambda_k + \lambda_l - \lambda)}{\sigma^2 - 2\sigma(\lambda_k + \lambda_l - \lambda) + \lambda_k^2 + \lambda_l^2 - \lambda(\lambda_k + \lambda_l) + \lambda_k\lambda_l}$$

The differentiation of λ' with respect to λ_k and λ_l together with,
$0 < \sigma < \lambda_1 \leq \lambda_k \leq \lambda_i < \lambda < \lambda_{i+1} \leq \lambda_l \leq \lambda_n$

$$\frac{\partial}{\partial\lambda_k}\lambda'\,(\lambda_k, \lambda_l, \lambda, \sigma) = \frac{[2(\lambda_k - \sigma) + \lambda_l - \lambda]\,(\lambda_l - \lambda)(\lambda_l - \sigma)^2}{(\sigma^2 + 2\sigma(\lambda - \lambda_k - \lambda_l) + \lambda_k^2 - \lambda_k\lambda + \lambda_k\lambda_l - \lambda_l\lambda + \lambda_l^2)^2} > 0$$

$$\frac{\partial}{\partial\lambda_l}\lambda'\,(\lambda_k, \lambda_l, \lambda, \sigma) = \frac{[2(\sigma - \lambda_l) + \lambda - \lambda_k]\,(\lambda - \lambda_k)(\lambda_k - \sigma)^2}{(\sigma^2 + 2\sigma(\lambda - \lambda_k - \lambda_l) + \lambda_k^2 - \lambda_k\lambda + \lambda_k\lambda_l - \lambda_l\lambda + \lambda_l^2)^2} > 0$$

Hence $\lambda'\,(\lambda_k, \lambda_l, \lambda, \sigma)$ takes its maximum in $\lambda'\,(\lambda_i, \lambda_i, \lambda, \sigma)$, whereas its minimum is taken in $\lambda'\,(\lambda_1, \lambda_n, \lambda, \sigma)$, i.e.

$$\lambda'\,(\lambda_1, \lambda_n, \lambda, \sigma), \leq \lambda\,((\wedge - \sigma I)^{-1}v) \leq \lambda'\,(\lambda_i, \lambda_{i+1}, \lambda, \sigma).$$

Reformulation of $(\lambda'\,(\lambda_1, \lambda_n, \lambda, \sigma), \leq \lambda\,((\wedge - \sigma I)^{-1}v) \leq \lambda'\,(\lambda_i, \lambda_{i+1}, \lambda, \sigma)$. using

$$\lambda' = \lambda'\,(\lambda_k, \lambda_l, \lambda, \sigma) = \frac{\sigma^2\lambda - 2\sigma\lambda_k\lambda_l + \lambda_k\lambda_l(\lambda_k + \lambda_l - \lambda)}{\sigma^2 - 2\sigma(\lambda_k + \lambda_l - \lambda) + \lambda_k^2 + \lambda_l^2 - \lambda(\lambda_k + \lambda_l) + \lambda_k\lambda_l}\,)\ \text{yields}$$

$$\frac{\lambda_1 + \lambda_n R_{1,n}(\lambda)}{1 + R_{1,n}(\lambda)} \leq \lambda' = \lambda\,((\wedge - \sigma I)^{-1}v) \leq \frac{\lambda_i + \lambda_{i+1}R_{i,i+1}(\lambda)}{1 + R_{i,i+1}(\lambda)}$$

with

$$R_{i,i+1}(\lambda) = \rho_{i,i+1}^2 \Delta_{i,i+1}(\lambda) = \left(\frac{\lambda_i - \sigma}{\lambda_{i+1} - \sigma}\right)^2 \frac{\lambda - \lambda_i}{\lambda_{i+1} - \lambda}$$

and

$$R_{1,n}(\lambda) = \rho_{1,n}^2 \Delta_{1,n}(\lambda) = \left(\frac{\lambda_i - \sigma}{\lambda_n - \sigma}\right)^2 \frac{\lambda - \lambda_1}{\lambda_n - \lambda}.$$

The right inequality of ($R_{i,i+1}(\lambda) = \rho_{i,i+1}^2 \Delta_{i,i+1}(\lambda) = \left(\dfrac{\lambda_i - \sigma}{\lambda_{i+1} - \sigma}\right)^2 \dfrac{\lambda - \lambda_i}{\lambda_{i+1} - \lambda}$) reads

$$\lambda' + \lambda' \rho_{i,i+1}^2 \Delta_{i,i+1}(\lambda) \leq \lambda_i + \lambda_i\, \rho_{i,i+1}^2 \Delta_{i,i+1}(\lambda),$$

from which ($\Delta_{i,i+1}(\lambda(x')) \leq (\rho_{i,i+1})^2 \Delta_{i,i+1}(\lambda).$) follows immediately. Reformulation of the left-hand inequality of ($\dfrac{\lambda_1 + \lambda_n R_{1,n}(\lambda)}{1 + R_{1,n}(\lambda)} \leq \lambda' = \lambda\left((\wedge - \sigma I)^{-1} v\right) \leq \dfrac{\lambda_i + \lambda_{i+1} R_{i,i+1}(\lambda)}{1 + R_{i,i+1}(\lambda)}$) proves
($(\rho_{1,n})^2\, \Delta_{i,n}(\lambda) \leq \Delta_{1,n}(\lambda(x')).$) analogously.

For $\sigma = 0$ the inequality ($\lambda'(\lambda_1, \lambda_n, \lambda, \sigma), \leq \lambda\left((\wedge - \sigma I)^{-1} v\right) \leq \lambda'(\lambda_i, \lambda_{i+1}, \lambda, \sigma).$) simply reads,

$$\frac{\lambda_1 \lambda_n (\lambda_1 + \lambda_n - \lambda)}{\lambda_n^2 - (\lambda - \lambda_1)(\lambda_1 - \lambda_n)} \leq \lambda' \leq \frac{\lambda_i \lambda_{i+1}(\lambda_i + \lambda_{i+1} - \lambda)}{\lambda_{i+1}^2 - (\lambda - \lambda_i)(\lambda_i - \lambda_{i+1})},$$

which proves ($\dfrac{1}{\lambda_1^{-1} + \lambda_n^{-1} - (\lambda_1 + \lambda_n - \lambda)^{-1}} \leq \lambda(x') \leq \dfrac{1}{\lambda_i^{-1} + \lambda_{i+1}^{-1} - (\lambda_1 + \lambda_{i+1} - \lambda)^{-1}}.$

The estimates ($\Delta_{i,i+1}(\lambda(x')) \leq (\rho_{i,i+1})^2 \Delta_{i,i+1}(\lambda).$) and ($(\rho_{1,n})^2\, \Delta_{i,n}(\lambda) \leq \Delta_{1,n}(\lambda(x')).$) are derived in the 2D invariant subspaces to either λ_i, λ_{i+1} or λ_1, λ_n and they are attained (by construction) exactly in these invariant subspaces. Therefore, ($\Delta_{i,i+1}(\lambda(x')) \leq (\rho_{i,i+1})^2 \Delta_{i,i+1}(\lambda).$ and $(\rho_{1,n})^2\, \Delta_{i,n}(\lambda) \leq \Delta_{1,n}(\lambda(x')).$) are each attained in a vector whose components are defined by ($v_k^2 = \dfrac{\lambda_l - \lambda}{\lambda_l - \lambda_k}$, and $v_l^2 = \dfrac{\lambda - \lambda_k}{\lambda_l - \lambda_k}.$) and whose Rayleigh quotients are given by ($\lambda' = \lambda'(\lambda_k, \lambda_l, \lambda, \sigma) = \dfrac{\sigma^2 \lambda - 2\sigma \lambda_k \lambda_l + \lambda_k \lambda_l (\lambda_k + \lambda_l - \lambda)}{\sigma^2 - 2\sigma(\lambda_k + \lambda_l - \lambda) + \lambda_k^2 + \lambda_l^2 - \lambda(\lambda_k + \lambda_l) + \lambda_k \lambda_l}.$)

References

- Farebrother, R.W. (1988), Linear Least Squares Computations, STATISTICS: Textbooks and Monographs, Marcel Dekker, ISBN 978-0-8247-7661-9

- Numerical-linear-algebra, dictionaries-thesauruses-pictures-and-press-releases: encyclopedia.com, Retrieved 22 June 2018

- Grcar, Joseph F. (2011a), "How ordinary elimination became Gaussian elimination", Historia Mathematica, 38 (2): 163–218, arXiv:0907.2397, doi:10.1016/j.hm.2010.06.003

- Gauss-jordan-elimination.html#gaussian-elimination-method: tutorvista.com, Retrieved 12 March 2018

- Golub, Gene H.; Van Loan, Charles F. (1996), Matrix Computations (3rd ed.), Johns Hopkins, ISBN 978-0-8018-5414-9

- Gaussian-elimination-method, gauss-jordan-elimination: tutorvista.com, Retrieved 29 May 2018

- Bunch, James R.; Hopcroft, John (1974), "Triangular factorization and inversion by fast matrix multiplication", Mathematics of Computation, 28 (125): 231–236, doi:10.2307/2005828, ISSN 0025-5718, JSTOR 2005828

- l-u-decomposition-system-linear-equations: geeksforgeeks.org, Retrieved 09 July 2018

Interpolation

In science and engineering, it is common to obtain a set of data points through sampling or experimentation. These represent the values of a function corresponding to values of an independent variable. To estimate the value of the function at on intermediate value of the variable, it is necessary to construct new data points in a process called interpolation. The aim of this chapter is to explore the fundamentals of interpolation, which includes the topics, polynomial interpolation, newton divided difference, LaGrange polynomial and cubic spline interpolation.

In mathematics, Interpolation is the determination or estimation of the value of $f(x)$, or a function of x, from certain known values of the function. If $x_0\ c\ ...<x_n$ and $y_0 = f(x_0),..., y_n = f(x_n)$ are known, and if $x_0 < x < x_n$, then the estimated value of $f(x)$ is said to be an interpolation. If $x < x_0$ or $x > x_n$, the estimated value of f(x) is said to be an extrapolation. If $x_0,..., x_n$ are given, along with corresponding values $y_0,..., y_n$, interpolation may be regarded as the determination of a function $y = f(x)$ whose graph passes through the n + 1 points, (x_i, y_i) for i = 0, 1,..., n. There are infinitely many such functions, but the simplest is a polynomial interpolation function. $y = p(x) = a_0 + a_1 x + ... + a_n x^n$ with constant $a_i's$ such that $p(x_i) = y_i$ for i = 0, ..., n. There is exactly one such interpolating polynomial of degree n or less. If the $x_i's$ are equally spaced, say by some factor h, then the following formula of Isaac Newton produces a polynomial function that fits the data:

$$f(x) = a_0 + \frac{a_1(x-x_0)}{h} + \frac{a_2(x-x_0)(x-x_1)}{2!h^2} + ... + \frac{a_n(x-x_0)(x-x_1)....(x-x_{n-1})}{n!h^n}$$

Polynomial approximation is useful even if the actual function $f(x)$ is not a polynomial, for the polynomial $p(x)$ often gives good estimates for other values of $f(x)$.

Piecewise Constant Interpolation

The easiest way to interpolate data is to do a piecewise constant interpolation.

We could choose the left endpoint of the interval:

$$P(x) = y_i, \quad x \in [x_i, x_{i+1}]$$

or the nearest point:

$$P(x) = y_i, \quad x \in \left[\frac{x_{i+1} + x_i}{2}, \frac{x_i + x_{i+1}}{2} \right]$$

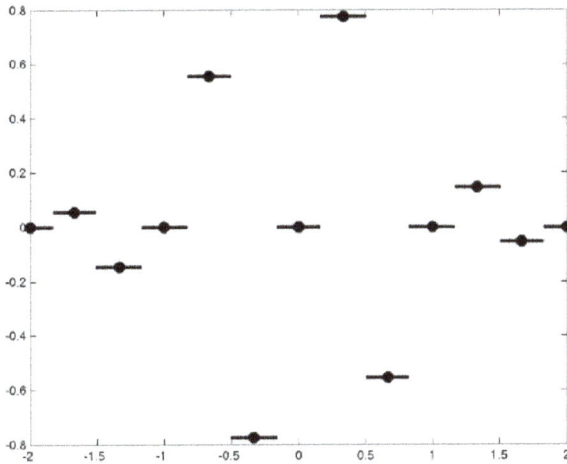

Choose the nearest value

Linear Interpolation

The linear interpolation is the straight line between the two points which are given by the coordinates (x_0, y_0) and (x_1, y_1). In the interval of (x_0, x_1) the value of x which gives the straight line and it is given from the equation for the value of y along.

$$\frac{y - y_0}{x - x_0} = \frac{y_1 - y_0}{x_1 - x_0}$$

It can be derived geometrically as follow,

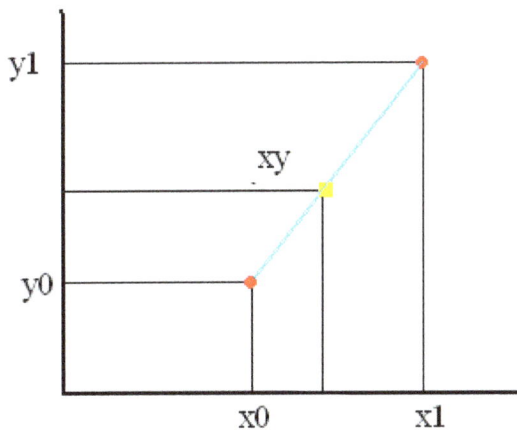

Solve the above for the unknown value of y,

$$y = y_0 + (x - x_0) \frac{y_1 - y_0}{x_1 - x_0}$$

In the interval (x_0, x_1), it is the linear interpolation formula.

Example: Problem for Linear Interpolation Formula

Some example problem for linear interpolation formula are,

Example:

Using linear interpolation formula, for the given coordinates of (1, 2) and (4, 5). Find the value for y when x = 2.

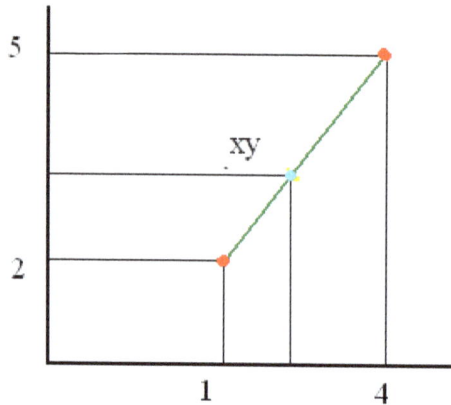

Solution:

Linear interpolation formula,

$$y = y_0 + (x - x_0) \frac{y_1 - y_0}{x_1 - x_0}$$

Given coordinate values are (1, 2) and (4, 5)

$(x_0, y_0))$ and (x_1, y_1) are (x_0, y_0) and (x_1, y_1).

$$y = 2 + (x - 1) \frac{(5 - 2)}{(4 - 1)}$$

$$y = 2 + x - 1\frac{3}{3}$$

$$y = 2 + x - 1(1)$$

$$y = x + 1.$$

Value of y when $x = 2$.

$$y = x + 1$$

$$x = 2$$

$$y = 2 + 1$$

$$y = 3.$$

Example:

Using the linear interpolation formula, find the equation for the given coordinates (6, 8) and (10, 16)

Solution:

Given coordinate values are (x_0, y_0) and (x_1, y_1) are (6, 8) and (10, 16).

$$y = y_0 + (x - x_0)\frac{y_1 - y_0}{x_1 - x_0}$$

$$y = 8 + (x - 6)\frac{16 - 8}{10 - 6}$$

$$y = 8 + (x - 6)(8/4)$$

$$y = 8 + (x - 6)(2)$$

$$y = 8 + 2x - 12$$

$$y = 2x - 12 + 8$$

$$y = 2x - 4$$

$$2x - y = 4 \text{ (or)}$$

$$2x - y - 4 = 0.$$

Spline Interpolation

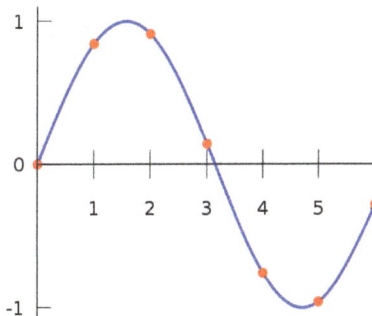

Plot of the data with spline interpolation applied

Remember that linear interpolation uses a linear function for each of intervals $[x_k, x_{k+1}]$. Spline interpolation uses low-degree polynomials in each of the intervals, and chooses the polynomial pieces such that they fit smoothly together. The resulting function is called a spline.

For instance, the natural cubic spline is piecewise cubic and twice continuously differentiable. Furthermore, its second derivative is zero at the end points. The natural cubic spline interpolating the points in the table above is given by,

$$f(x) = \begin{cases} -0.1522x^3 + 0.9937x, & \text{if } x \in [0,1], \\ -0.01258x^3 - 0.4189x^2 + 1.4126x - 0.1396, & \text{if } x \in [1,2], \\ 0.1403x^3 - 1.3359x^2 + 3.2467x - 1.3623, & \text{if } x \in [2,3], \\ 0.1579x^3 - 1.4945x^2 + 3.7225x - 1.8381, & \text{if } x \in [3,4], \\ 0.05375x^3 - 0.2450x^2 - 1.2756x + 4.8259, & \text{if } x \in [4,5], \\ -0.1871x^3 + 3.3673x^2 - 19.3370x + 34.9282, & \text{if } x \in [5,6]. \end{cases}$$

In this case we get $f(2.5) = 0.5972$.

Like polynomial interpolation, spline interpolation incurs a smaller error than linear interpolation and the interpolant is smoother. However, the interpolant is easier to evaluate than the high-degree polynomials used in polynomial interpolation. However, the global nature of the basis functions leads to ill-conditioning. This is completely mitigated by using splines of compact support.

Via Gaussian Processes

Gaussian process is a powerful non-linear interpolation tool. Many popular interpolation tools are actually equivalent to particular Gaussian processes. Gaussian processes can be used not only for fitting an interpolant that passes exactly through the given data points but also for regression, i.e., for fitting a curve through noisy data. In the geostatistics community Gaussian process regression is also known as Kriging.

Other forms

Other forms of interpolation can be constructed by picking a different class of interpolants. For instance, rational interpolation is interpolation by rational functions using Padé approximant, and trigonometric interpolation is interpolation by trigonometric polynomials using Fourier series. Another possibility is to use wavelets.

The Whittaker–Shannon interpolation formula can be used if the number of data points is infinite.

Sometimes, we know not only the value of the function that we want to interpolate, at some points, but also its derivative. This leads to Hermite interpolation problems.

When each data point is itself a function, it can be useful to see the interpolation problem as a partial advection problem between each data point. This idea leads to the displacement interpolation problem used in transportation theory.

In Higher Dimensions

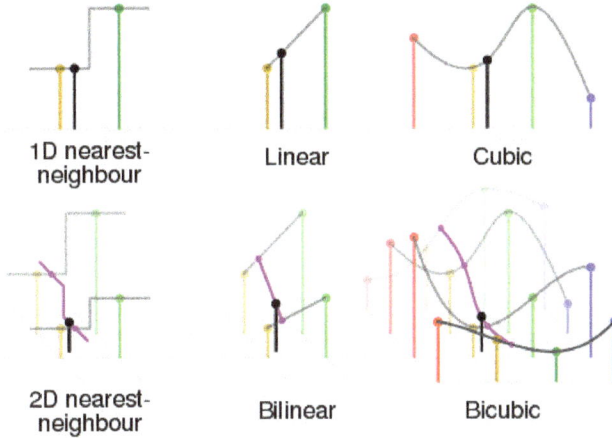

| 1D nearest-neighbour | Linear | Cubic |

| 2D nearest-neighbour | Bilinear | Bicubic |

Comparison of some 1- and 2-dimensional interpolations. Black and red/yellow/green/blue dots correspond to the interpolated point and neighbouring samples, respectively. Their heights above the ground correspond to their values

Multivariate interpolation is the interpolation of functions of more than one variable. Methods include bilinear interpolation and bicubic interpolation in two dimensions, and trilinear interpolation in three dimensions. They can be applied to gridded or scattered data.

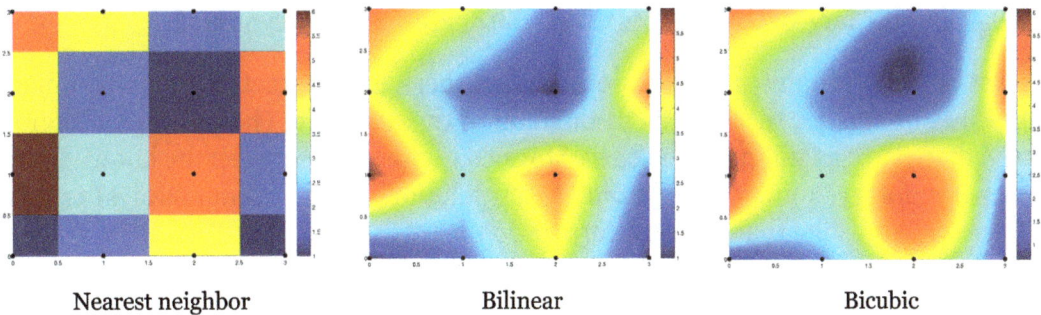

| Nearest neighbor | Bilinear | Bicubic |

In Digital Signal Processing

In the domain of digital signal processing, the term interpolation refers to the process of converting a sampled digital signal (such as a sampled audio signal) to that of a higher sampling rate (upsampling) using various digital filtering techniques (e.g., convolution with a frequency-limited impulse signal). In this application there is a specific requirement that the harmonic content of the original signal be preserved without creating aliased harmonic content of the original signal above the original Nyquist limit of the signal (i.e., above fs/2 of the original signal sample rate). An early and fairly elementary discussion on this subject can be found in Rabiner and Crochiere's book *Multirate Digital Signal Processing*.

Polynomial Interpolation

Polynomial interpolation is a method of estimating values between known data points. When graphical data contains a gap, but data is available on either side of the gap or at a few specific points within the gap, an estimate of values within the gap can be made by interpolation.

The simplest method of interpolation is to draw straight lines between the known data points and consider the function as the combination of those straight lines. This method, called linear interpolation, usually introduces considerable error. A more precise approach uses a polynomial function to connect the points. A polynomial is a mathematical expression comprising a sum of terms, each term including a variable or variables raised to a power and multiplied by a coefficient. The simplest polynomials have one variable. Polynomials can exist in factored form or written out in full. For example:

$$(x-4)(x+2)(x+10)$$

$$x^2 + 2x + 1$$

$$3y^3 - 8y^2 + 4y - 2$$

The value of the largest exponent is called the degree of the polynomial.

If a set of data contains n known points, then there exists exactly one polynomial of degree $n-1$ or smaller that passes through all of those points. The polynomial's graph can be thought of as "filling in the curve" to account for data between the known points. This methodology, known as polynomial interpolation, often (but not always) provides more accurate results than linear interpolation.

The main problem with polynomial interpolation arises from the fact that even when a certain polynomial function passes through all known data points, the resulting graph might not reflect the actual state of affairs. It is possible that a polynomial function, although accurate at specific points, will differ wildly from the true values at some regions between those points. This problem most often arises when "spikes" or "dips" occur in a graph, reflecting unusual or unexpected events in a real-world situation. Such anomalies are not reflected in the simple polynomial function which, even though it might make perfect mathematical sense, cannot take into account the chaotic nature of events in the physical universe.

Example: Let f be a given function defined on an interval [a,b], and let $(x_i)_{i=0}^n$ be n +1 distinct points in [a,b]. The polynomial interpolation problem is to find a polynomial $p_n = (f; x_0, \ldots, x_n)$ of degree n that matches f at the given points,

$$p_n(x_i) = f(x_i), \text{ for } i = 0, 1, \ldots, n.$$

The points $(x_i)_{i=0}^n$ are called interpolation points, the conditions (9.17) are called the interpolation conditions, and the polynomial $p_n = (f; x_0, \ldots, x_n)$ is called a polynomial interpolant.

Example: Let $(x_i)_{i=0}^n$ be n+1 distinct real numbers. The Newton form of a polynomial of degree n is an expression in the form,

$$p_n(x) = c_0 + c_1(x - x_0) + c_2(x - x_0)(x - x_1) + \ldots + c_n(x - x_0)(x - x_1)\cdots(x - x_n - 1).$$

The advantage of the Newton form will become evident when we consider some examples.

Example: (Newton form for $n = 0$). Suppose we have only one interpolation point x_0. Then the Newton form is just $p_0(x) = c_0$. To interpolate f at x_0 we simply have to choose $c_0 = f(x_0)$,

$$p_0(x) = f(x_0).$$

Example: (Newton form for $n = 1$). With two points x_0 and x_1 the Newton form is $p_1(x) = c_0 + c_1(x - x_0)$. Interpolation at x0 means that $f(x_0) = p_1(x_0) = c_0$ and interpolation at x_1,

$$f(x_1) = p_1(x_1) = f(x_0) + c_1(x_1 - x_0),$$

which means that,

$$c_0 = f(x_0), \ c_1 = \frac{f(x_1) - f(x_0)}{x_1 - x_0}.$$

We note that c_0 remains the same as in the case $n = 0$.

Example: (Newton form for $n = 2$). We add another point and consider interpolation with a quadratic polynomial,

$$p_2(x) = c_0 + c_1(x - x_0) + c_2(x - x_0)(x - x_1).$$

at the three points x_0, x_1, x_2. Interpolation at x_0 and x_1 gives the equations,

$$f(x_0)\, p_2(x_0) = c_0,$$
$$f(x_1)\, p_2(x_1) = c_0 + c_1 + (x_1 - x_0),$$

which we note are the same equations as we solved in the case $n = 1$. From the third condition,

$$f(x_2)\, p(x_2) = c_0 + c_1(x_2 - x_0) + c_2(x_2 - x_0)(x_2 - x_1),$$

we obtain,

$$c_2 = \frac{f(x_2) - f(x_0) - \dfrac{f(x_1) - f(x_0)}{x_1 - x_0}(x_2 - x_0)}{(x_2 - x_0)(x_2 - x_1)}.$$

Playing around a bit with this expression one finds that it can also be written as,

$$c_2 = \frac{\dfrac{f(x_2) - f(x_1)}{x_2 - x_1} - \dfrac{f(x_1) - f(x_0)}{x_1 - x_0}}{x_2 - x_0}.$$

It is easy to see that what happened in the quadratic case happens in the general case: The equation that results from the interpolation condition at x_k involves only the points $(x_0, f(x_0)), (x_1, f(x_1)), \ldots, (x_k, f(x_k))$. If we write down all the equations we find,

$$f(x_0) = c_0,$$
$$f(x_1) = c_0, + c_1\,(x_1 - x_0),$$
$$f(x_2) = c_0, + c_1\,(x_2 - x_0) + c_2(x_2 - x_0)(x_2 - x_1),$$
$$\vdots$$
$$f(x_k) = c_0 + c_1(x_k - x_0) + c_2(x_k - x_0)(x_k - x_1) + \cdots$$
$$+ c_{k-1}(x_k - x_0)\cdots(x_k - x_k - 2) + c_k\,(x_k - x_0)\cdots(x_k - x_k - 1).$$

This is an example of a triangular system where each new equation introduces one new variable and one new point. This means that each coefficient c_k depends on the data $(x_0, f(x_0)), (x_1, f(x_1)), \ldots, (x_k, f(x_k))$, and the following theorem is immediate.

Theorem: Let f be a given function and $x_0, \ldots, x_n$ given and distinct interpolation points. There is a unique polynomial of degree n which interpolates f at these points. If the interpolating polynomial is expressed in Newton form,

$$p_n(x) = c_0 + c_1(x - x_0) + \ldots + c_n(x - x_0)(x - x_1)\ldots(x - x_{n-1}),$$

then c_k depends only on $(x_0, f(x_0)), (x_1, f(x_1)), \ldots, (x_k, f(x_k))$ which is indicated by the notation,

$$c_k = f[x_0, \ldots, x_k]$$

For $k = 0, 1, \ldots, n$. The interpolating polynomials p_n and p_{n-1} are related by,

$$p_n(x) = p_{n-1}(x) + f[x_0, \ldots, x_n](x - x_0) \ldots (x - x_{n-1}).$$

Note that with the notation for the coefficients, the interpolation formula becomes,

$$p_n(x) = f[x_0] + f[x_0, x_1](x - x_0) + \ldots + f[x_0, \ldots, x_n](x - x_0) \ldots (x - x_{n-1}).$$

The coefficients (c_k) can be computed from the triangular system equations. A very useful feature of the Newton form is that an additional interpolation point just adds an additional equation to triangular system equation, with the earlier equations being unaffected by the new condition. This means that the additional coefficient can be computed from this new equation, while the other coefficients remain the same as in the case with one point less.

Example: Suppose we have the four points $x_i = i$, for $i = 0, \ldots, 3$ and we want to interpolate the function $\sqrt{x}$ at these points. In this case the Newton form becomes,

$$p_3(x) = c_0 + c_1 x + c_2 x(x - 1) + c_3 x(x - 1)(x - 2).$$

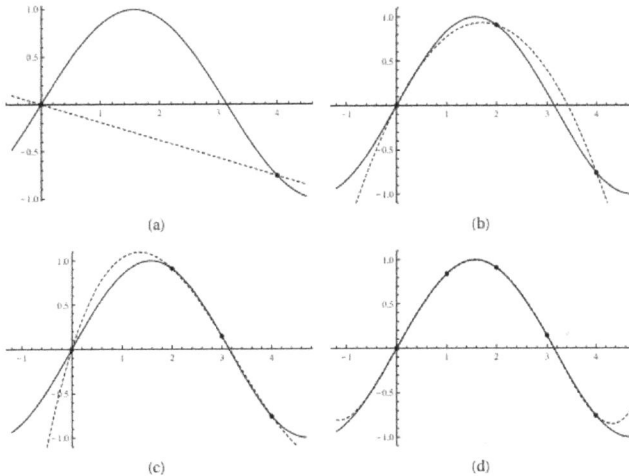

Figure: Interpolation of sin x with a line (a), a parabola (b), a cubic (c), and a quartic polynomial (d)

The interpolation conditions become,

$$0 = c_0,$$
$$1 = c_0 + c_1,$$
$$\sqrt{2} = c_0 + 2c_1 + 2c_2,$$
$$\sqrt{3} = c_0 + 3c_1 + 6c_2 + 6c_3.$$

Not surprisingly, the equations are triangular and we find,

$$c_0 = 0, c_1 = 1, c_2 = -(1 - \sqrt{2}/2), c_3 = (3 + \sqrt{3} - 3\sqrt{2})/6$$

Non-vandermonde Solutions

We are trying to construct our unique interpolation polynomial in the vector space Π_n of polynomials of degree n. When using a monomial basis for Π_n we have to solve the Vandermonde matrix to construct the coefficients a_k for the interpolation polynomial. This can be a very costly operation (as counted in clock cycles of a computer trying to do the job). By choosing another basis for Π_n we can simplify the calculation of the co-efficients but then we have to do additional calculations when we want to express the interpolation polynomial in terms of a monomial basis.

One method is to write the interpolation polynomial in the Newton form and use the method of divided differences to construct the coefficients, e.g. Neville's algorithm. The cost is $O(n^2)$ operations, while Gaussian elimination costs $O(n^3)$ operations. Further-more, you only need to do $O(n)$ extra work if an extra point is added to the data set, while for the other methods, you have to redo the whole computation.

Another method is to use the Lagrange form of the interpolation polynomial. The re-sulting formula immediately shows that the interpolation polynomial exists under the conditions stated in the above theorem. Lagrange formula is to be preferred to Vander-monde formula when we are not interested in computing the coefficients of the polyno-mial, but in computing the value of $p(x)$ in a given x not in the original data set. In this case, we can reduce complexity to $O(n^2)$.

The Bernstein form was used in a constructive proof of the Weierstrass approximation theorem by Bernstein and has nowadays gained great importance in computer graphics in the form of Bézier curves.

Linear Combination of the given Values

The Lagrange form of the interpolating polynomial is a linear combination of the given values. In many scenarios, an efficient and convenient polynomial interpolation is a linear combination of the given values, using previously known coefficients. Given a set of $k+1$ data points,

$$(x_0, y_0), \ldots, (x_j, y_j), \ldots, (x_k, y_k)$$

where each data point is a (position, value) pair and where no two positions x_j are the same, the interpolation polynomial in the Lagrange form is a linear combination,

$$y(x) := \sum_{j=0}^{k} y_j c_j(x)$$

of the given values y_j with each coefficient $c_j(x)$ given by evaluating the corresponding Lagrange basis polynomial using the given $k+1$ positions x_j.

$$c_j(x) = \ell_j(x, x_0, x_1, \ldots, x_k) := \prod_{\substack{0 \le m \le k \\ m \ne j}} \frac{x - x_m}{x_j - x_m} = \frac{(x - x_0)}{(x_j - x_0)} \cdots \frac{(x - x_{j-1})}{(x_j - x_{j-1})} \frac{(x - x_{j+1})}{(x_j - x_{j+1})} \cdots \frac{(x - x_k)}{(x_j - x_k)}.$$

Each coefficient $c_j(x)$ in the linear combination depends on the given positions x_j and the desired position x, but not on the given values y_j. For each coefficient, inserting the values of the given positions x_j and simplifying yields an expression $c_j(x)$, which depends only on x. Thus the same coefficient expressions $c_j(x)$ can be used in a polynomial interpolation of a given second set of $k+1$ data points,

$$(x_0, v_0), \ldots, (x_j, v_j), \ldots, (x_k, v_k)$$

at the same given positions x_j, where the second given values v_j differ from the first given values y_j. Using the same coefficient expressions $c_j(x)$ as for the first set of data points, the interpolation polynomial of the second set of data points is the linear combination,

$$v(x) := \sum_{j=0}^{k} v_j c_j(x).$$

For each coefficient $c_j(x)$ in the linear combination, the expression resulting from the Lagrange basis polynomial $\ell_j(x, x_0, x_1, \ldots, x_k)$ only depends on the relative spaces between the given positions, not on the individual value of any position. Thus the same coefficient expressions $c_j(x)$ can be used in a polynomial interpolation of a given third set of $k+1$ data points,

$$(t_0, w_0), \ldots, (t_j, w_j), \ldots, (t_k, w_k)$$

where each position t_j is related to the corresponding position x_j in the first set by $t_i = a * x_i + b$ and the desired positions are related by $t = a * x + b$, for a constant scaling factor a and a constant shift b for all positions. Using the same coefficient expressions $c_j(t)$ as for the first set of data points, the interpolation polynomial of the third set of data points is the linear combination,

$$(t) := \sum_{j=0}^{k} w_j c_j(t).$$

In many applications of polynomial interpolation, the given set of $k+1$ data points is at equally spaced positions. In this case, it can be convenient to define the x-axis of the positions such that $x_i = i$. For example, a given set of 3 equally-spaced data points $(x_0, y_0), (x_1, y_1), (x_2, y_2)$ is then $(0, y_0), (1, y_1), (2, y_2)$.

The interpolation polynomial in the Lagrange form is the linear combination,

$$y(x) := \sum_{j=0}^{2} y_j c_j(x) = y_0 \frac{(x-1)(x-2)}{(0-1)(0-2)} + y_1 \frac{(x-0)(x-2)}{(1-0)(1-2)} + y_2 \frac{(x-0)(x-1)}{(2-0)(2-1)}$$

$$= y_0 \frac{(x-1)(x-2)}{2} + y_1 \frac{(x-0)(x-2)}{-1} + y_2 \frac{(x-0)(x-1)}{2}.$$

This quadratic interpolation is valid for any position x, near or far from the given positions. So, given 3 equally-spaced data points at $x = 0, 1, 2$ defining a quadratic polynomial, at an example desired position $x = 1.5$, the interpolated value after simplification is given by,

$$y(1.5) = y_{1.5} = (-y_0 + 6y_1 + 3y_2)/8$$

This is a quadratic interpolation typically used in the Multigrid method.

Again given 3 equally-spaced data points at $x = 0, 1, 2$ defining a quadratic polynomial, at the next equally spaced position $x = 3$, the interpolated value after simplification is given by,

$$y(3) = y_3 = y_0 - 3y_1 + 3y_2$$

In the above polynomial interpolations using a linear combination of the given values, the coefficients were determined using the Lagrange method. In some scenarios, the coefficients can be more easily determined using other methods. Examples follow.

According to the method of finite differences, for any polynomial of degree d or less, any sequence of $d+2$ values at equally spaced positions has a $(d+1)$th difference exactly equal to 0. The element s_{d+1} of the Binomial transform is such a $(d+1)$th difference. This area is surveyed here. The binomial transform, T, of a sequence of values $\{v_n\}$, is the sequence $\{s_n\}$ defined by,

$$s_n = \sum_{k=0}^{n} (-1)^k \binom{n}{k} v_k.$$

Ignoring the sign term $(-1)^k$, the $n+1$ coefficients of the element s_n are the respective $n+1$ elements of the row n of Pascal's Triangle. The triangle of binomial transform coefficients is like Pascal's triangle. The entry in the nth row and kth column of the BTC triangle is $(-1)^k \binom{n}{k}$ for any non-negative integer n and any integer k between 0 and n. This results in the following example rows $n = 0$ through $n = 7$, top to bottom, for the BTC triangle:

1   // Row n = 0.

1 -1   // Row n = 1 or d = 0.

1 -2 1   // Row n = 2 or d = 1.

1 -3 3 -1   // Row n = 3 or d = 2.

1 -4 6 -4 1   // Row n = 4 or d = 3.

1 -5 10 -10 5 -1   // Row n = 5 or d = 4.

1 -6 15 -20 15 -6 1   // Row n = 6 or d = 5.

1 -7 21 -35 35 -21 7 -1   // Row n = 7 or d = 6.

For convenience, each row n of the above example BTC triangle also has a label $d = n-1$. Thus for any polynomial of degree d or less, any sequence of $d+2$ values at equally spaced positions has a linear combination result of 0, when using the $d+2$ elements of row d as the corresponding linear coefficients.

For example, 4 equally spaced data points of a quadratic polynomial obey the linear equation given by row $d = 2$ of the BTC triangle.

$$0 = y_0 - 3y_1 + 3y_2 - y_3$$

This is the same linear equation as obtained above using the Lagrange method.

The BTC triangle can also be used to derive other polynomial interpolations. For example, the above quadratic interpolation,

$$y(1.5) = y_{1.5} = (-y_0 + 6y_1 + 3y_2)/8$$

can be derived in 3 simple steps as follows. The equally spaced points of a quadratic polynomial obey the rows of the BTC triangle with $d = 2$ or higher. First, the row $d = 3$ spans the given and desired data points $y_0, y_1, y_{1.5}, y_2$ with the linear equation,

$$0 = 1y_0 - 4y_{0.5} + 6y_1 - 4y_{1.5} + 1y_2$$

Second, the unwanted data point $y_{0.5}$ is replaced by an expression in terms of wanted data points. The row $d = 2$ provides a linear equation with a term $1y_{0.5}$, which results in a term $4y_{0.5}$ by multiplying the linear equation by 4.

$$0 = 1y_{0.5} - 3y_1 + 3y_{1.5} - 1y_2 = 4y_{0.5} - 12y_1 + 12y_{1.5} - 4y_2$$

Third, the above two linear equations are added to yield a linear equation equivalent to the above quadratic interpolation for $y_{1.5}$.

$$0 = (1+0)y_0 + (-4+4)y_{0.5} + (6-12)y_1 + (-4+12)y_{1.5} + (1-4)y_2 = y_0 - 6y_1 + 8y_{1.5} - 3y_2$$

Similar to other uses of linear equations, the above derivation scales and adds vectors of coefficients. In polynomial interpolation as a linear combination of values, the elements of a vector correspond to a contiguous sequence of regularly spaced positions. The p non-zero elements of a vector are the p coefficients in a linear equation obeyed by any sequence of p data points from any degree d polynomial on any regularly spaced grid, where d is noted by the subscript of the vector. For any vector of coefficients, the subscript obeys $d \le p-2$. When adding vectors with various subscript values, the lowest subscript applies for the resulting vector. So, starting with the vector of row $d = 3$ and the vector of row $d = 2$ of the BTC triangle, the above quadratic interpolation for $y_{1.5}$ is derived by the vector calculation,

$$(1,-4,6,-4,1)_3 + 4*(0,1,-3,3,-1)_2 = (1,0,-6,+8,-3)_2$$

Similarly, the cubic interpolation typical in the Multigrid method,

$$y_{1.5} = (-y_0 + 9y_1 + 9y_2 - y_3)/16,$$

can be derived by a vector calculation starting with the vector of row $d = 5$ and the vector of row $d = 3$ of the BTC triangle.

$$(1,-6,15,-20,15,-6,1)_5 + 6*(0,1,-4,6,-4,1,0)_3 = (1,0,-9,16,-9,0,1)_3$$

Interpolation Error

When interpolating a given function f by a polynomial of degree n at the nodes $x_0,...,x_n$ we get the error,

$$f(x) - p_n(x) = f[x_0,...,x_n,x]\prod_{i=0}^{n}(x - x_i)$$

where

$$f[x,...,x,x]$$

is the notation for divided differences.

If f is $n + 1$ times continuously differentiable on a closed interval I and $p_n(x)$ is a polynomial of degree at most n that interpolates f at $n + 1$ distinct points $\{x_i\}$ ($i=0,1,...,n$) in that interval, then for each x in the interval there exists ξ in that interval such that,

$$f(x) - p_n(x) = \frac{f^{(n+1)}(\xi)}{(n+1)!}\prod_{i=0}^{n}(x - x_i).$$

The above error bound suggests choosing the interpolation points x_i such that the product $\left|\prod(x - x_i)\right|$, is as small as possible. The Chebyshev nodes achieve this.

Proof

Set the error term as,

$$R_n(x) = f(x) - p_n(x)$$

and set up an auxiliary function:

$$Y(t) = R_n(t) - \frac{R_n(x)}{W(x)} W(t)$$

where

$$W(t) = \prod_{i=0}^{n} (t - x_i).$$

Since x_i are roots of $R_n(t)$ and $W(t)$, we have $Y(x) = Y(x_i) = 0$, which means Y has at least $n + 2$ roots. From Rolle's theorem, $Y'(t)$ has at least $n + 1$ roots, then $Y^{(n+1)}(t)$ has at least one root ξ, where ξ is in the interval I.

So we can get,

$$Y^{(n+1)}(t) = R_n^{(n+1)}(t) - \frac{R_n(x)}{W(x)} (n+1)!$$

Since $p_n(x)$ is a polynomial of degree at most n, then,

$$R_n^{(n+1)}(t) = f^{(n+1)}(t)$$

Thus

$$Y^{(n+1)}(t) = f^{(n+1)}(t) - \frac{R_n(x)}{W(x)} (n+1)!$$

Since ξ is the root of $Y^{(n+1)}(t)$, so

$$Y^{(n+1)}(\xi) = f^{(n+1)}(\xi) - \frac{R_n(x)}{W(x)} (n+1)! = 0$$

Therefore,

$$R_n(x) = f(x) - p_n(x) = \frac{f^{(n+1)}(\xi)}{(n+1)!} \prod_{i=0}^{n} (x - x_i).$$

Thus the remainder term in the Lagrange form of the Taylor theorem is a special case of interpolation error when all interpolation nodes x_i are identical. Note that the error will be zero when $x = x_i$ for any i. Thus, the maximum error will occur at some point in the interval between two successive nodes.

For Equally Spaced Intervals

In the case of equally spaced interpolation nodes where $x_i = a + ih$, for $i = 0, 1, \ldots, n$, and where $h = (b-a)/n$, the product term in the interpolation error formula can be bound as,

$$\left| \prod_{i=0}^{n} (x - x_i) \right| = \prod_{i=0}^{n} |x - x_i| \le \frac{n!}{4} h^{n+1}.$$

Thus the error bound can be given as,

$$|R_n(x)| \le \frac{h^{n+1}}{4(n+1)} \max_{\xi \in [a,b]} \left| f^{(n+1)}(\xi) \right|$$

However, this assumes that $f^{(n+1)}(\xi)$ is dominated by h^{n+1}, i.e. $f^{(n+1)}(\xi) h^{n+1} \ll 1$. In several cases, this is not true and the error actually increases as $n \to \infty$.

Lebesgue Constants

We fix the interpolation nodes $x_0, \ldots, x_n$ and an interval $[a, b]$ containing all the interpolation nodes. The process of interpolation maps the function f to a polynomial p. This defines a mapping X from the space $C([a, b])$ of all continuous functions on $[a, b]$ to itself. The map X is linear and it is a projection on the subspace Π_n of polynomials of degree n or less.

The Lebesgue constant L is defined as the operator norm of X. One has (a special case of Lebesgue's lemma):

$$\| f - X(f) \| \le (L+1) \| f - p^* \|.$$

In other words, the interpolation polynomial is at most a factor $(L + 1)$ worse than the best possible approximation. This suggests that we look for a set of interpolation nodes that makes L small. In particular, we have for Chebyshev nodes:

$$L \le \frac{2}{\pi} \log(n+1) + 1.$$

We conclude again that Chebyshev nodes are a very good choice for polynomial interpolation, as the growth in n is exponential for equidistant nodes. However, those nodes are not optimal.

Convergence Properties

It is natural to ask, for which classes of functions and for which interpolation nodes the sequence of interpolating polynomials converges to the interpolated function as $n \to \infty$? Convergence may be understood in different ways, e.g. pointwise, uniform or in some integral norm.

The situation is rather bad for equidistant nodes, in that uniform convergence is not even guaranteed for infinitely differentiable functions. One classical example, due to Carl Runge, is the function $f(x) = 1 / (1 + x^2)$ on the interval $[-5, 5]$. The interpolation error $||f - p_n||_\infty$ grows without bound as $n \to \infty$. Another example is the function $f(x) = |x|$ on the interval $[-1, 1]$, for which the interpolating polynomials do not even converge pointwise except at the three points $x = \pm 1, 0$.

One might think that better convergence properties may be obtained by choosing different interpolation nodes. The following result seems to give a rather encouraging answer:

> Theorem: For any function f(x) continuous on an interval [a,b] there exists a table of nodes for which the sequence of interpolating polynomials $p_n(x)$ converges to f(x) uniformly on [a,b].

Proof. It's clear that the sequence of polynomials of best approximation $p_n^*(x)$ converges to $f(x)$ uniformly (due to Weierstrass approximation theorem). Now we have only to show that each $p_n^*(x)$ may be obtained by means of interpolation on certain nodes. But this is true due to a special property of polynomials of best approximation known from the equioscillation theorem. Specifically, we know that such polynomials should intersect $f(x)$ at least $n + 1$ times. Choosing the points of intersection as interpolation nodes we obtain the interpolating polynomial coinciding with the best approximation polynomial.

The defect of this method, however, is that interpolation nodes should be calculated anew for each new function $f(x)$, but the algorithm is hard to be implemented numerically. Does there exist a single table of nodes for which the sequence of interpolating polynomials converge to any continuous function $f(x)$? The answer is unfortunately negative:

> Theorem. For any table of nodes there is a continuous function f(x) on an interval [a, b] for which the sequence of interpolating polynomials diverges on [a,b].

The proof essentially uses the lower bound estimation of the Lebesgue constant, which we defined above to be the operator norm of X_n (where X_n is the projection operator on Π_n). Now we seek a table of nodes for which

$$\lim_{n \to \infty} X_n f = f, \text{ for every } f \in C([a,b]).$$

Due to the Banach–Steinhaus theorem, this is only possible when norms of X_n are uniformly bounded, which cannot be true since we know that,

$$||X_n|| \geq \frac{2}{\pi} \log(n+1) + C.$$

For example, if equidistant points are chosen as interpolation nodes, the function from Runge's phenomenon demonstrates divergence of such interpolation. Note that this function is not only continuous but even infinitely times differentiable on [−1, 1]. For better Chebyshev nodes, however, such an example is much harder to find due to the following result:

> Theorem. For every absolutely continuous function on [−1, 1] the sequence of interpolating polynomials constructed on Chebyshev nodes converges to f(x) uniformly.

Applications

Polynomials can be used to approximate complicated curves, for example, the shapes of letters in typography, given a few points. A relevant application is the evaluation of the natural logarithm and trigonometric functions: pick a few known data points, create a lookup table, and interpolate between those data points. This results in significantly faster computations. Polynomial interpolation also forms the basis for algorithms in numerical quadrature and numerical ordinary differential equations and Secure Multi Party Computation, Secret Sharing schemes.

Polynomial interpolation is also essential to perform sub-quadratic multiplication and squaring such as Karatsuba multiplication and Toom–Cook multiplication, where an interpolation through points on a polynomial which defines the product yields the product itself. For example, given $a = f(x) = a_0 x^0 + a_1 x^1 + \ldots$ and $b = g(x) = b_0 x^0 + b_1 x^1 + \ldots$, the product ab is equivalent to $W(x) = f(x)g(x)$. Finding points along $W(x)$ by substituting x for small values in $f(x)$ and $g(x)$ yields points on the curve. Interpolation based on those points will yield the terms of $W(x)$ and subsequently the product ab. In the case of Karatsuba multiplication this technique is substantially faster than quadratic multiplication, even for modest-sized inputs. This is especially true when implemented in parallel hardware.

Newton Divided Difference

Newton's divided difference interpolation formula is a interpolation technique used when the interval difference is not same for all sequence of values.

Let us assume that the function $f(x)$ is linear then we have $\dfrac{f(x_i) - f(x_j)}{(x_i - x_j)}$

where x_i and x_j are any two tabular points, is independent of x_i and x_j. This ratio is called the first divided difference of $f(x)$ relative to x_i and x_j and is denoted by $f[x_i, x_j]$. That is,

$$f[x_i, x_j] = \frac{f(x_i) - f(x_j)}{(x_i - x_j)} \, f[x_j, x_i]$$

Since the ratio is independent of x_i and x_j we can write $f[x_0, x_1] = f[x_0, x_1]$

$$\frac{f(x) - f(x_0)}{(x - x_0)} = f[x_0, x_1]$$

$$f(x) = f(x_0) + (x - x_0) \, f[x_0, x_1]$$

$$= \frac{1}{x - x_0} \begin{vmatrix} f(x_0) & x_0 - x \\ f(x_1) & x_1 - x \end{vmatrix} = \frac{f_1 - f_0}{x_1 - x_0} x + \frac{f_0 x_1 - f_1 x_0}{x_1 - x_0}$$

So if $f(x)$ is approximated with a linear polynomial then the function value at any point x can be calculated by using,

$$f(x) \cong p_1(x) = f(x_0) + (x - x_1) \, f[x_0, x_1]$$

where $f[x_0, x_1]$ is the first divided difference of f relative to x_0 and x_1.

Similarly if $f(x)$ is a second degree polynomial then the secant slope defined above is not constant but a linear function of x. Hence we have,

$$\frac{f[x_1, x_2] - f[x_0, x_1]}{x_2 - x_0}$$

is independent of x_0, x_1 and x_2. This ratio is defined as second divided difference of f relative to x_0, x_1 and x_2. The second divided difference are denoted as,

$$f[x_0, x_1, x_2] = \frac{f[x_1, x_2] - f[x_0, x_1]}{x_2 - x_0}$$

Now again since $f[x_0, x_1, x_2]$ is independent of x_0, x_1 and x_2 we have,

$$f[x_1, x_0, x] = f[x_0, x_1, x_2]$$

$$\frac{f[x_0, x] - f[x_1, x_0]}{x - x_1} = f[x_0, x_1, x_2]$$

$$f[x_0, x_1] = f[x_0, x_1] + (x - x_1) \, f[x_0, x_1, x_2]$$

$$\frac{f(x) - f(x_0)}{(x - x_0)} = f[x_0, x_1] + (x - x_1) f[x_0, x_1, x_2]$$

$$f(x) = f(x_0) + (x - x_0) f[x_0, x_1] + (x - x_0)(x - x_1) f[x_0, x_1, x_2]$$

This is equivalent to the second degree polynomial approximation passing through three data points,

x_0	x_1	x_2
f_0	f_1	f_2

So whenever $f(x)$ is approximated with a second degree polynomial, the value of f(x) at any point x can be computed using the above polynomial.

In the same way if we define recursively k^{th} divided difference by the relation,

$$f[x_0, x_1, \ldots x_k] = \frac{f[x_1, x_2, \ldots x_k] - f[x_0, x_1, \ldots x_{k-1}]}{x_k - x_0}$$

The kth degree polynomial approximation to f(x) can be written as,

$$f(x) = f(x_0) + (x - x_0) f[x_0, x_1] + (x - x_0)(x - x_1) f[x_0, x_1, x_2]$$
$$+ \ldots + (x - x_0)(x - x_1) \ldots (x - x_{k-1}) f[x_0, x_1, \ldots x_k].$$

This formula is called Newton's Divided Difference Formula. Once we have the divided differences of the function f relative to the tabular points then we can use the above formula to compute f(x) at any non tabular point.

Computing divided differences using divided difference table: Let us consider the points $(x_1, f_1), (x_2, f_2), (x_3, f_3)$ and (x_4, f_4) where x_1, x_2, x_3 and x_4 are not necessarily equi-distant points then the divided difference table can be written as,

x_1		$f[x_1, x_j]$	$f[x_i, x_j, x_k]$	$f[x_i, x_j, x_k, x_l]$
x_1	f_1			
		$f[x_1, x_2] = \dfrac{f_2 - f_1}{x_2 - x_1}$		
x_2	f_2		$f[x_1, x_2, x_3] = \dfrac{f[x_2, x_3] - f[x_1, x_2]}{x_3 - x_1}$	

		$f[x_2,x_3]=\dfrac{f_3-f_2}{x_3-x_2}$		$f[x_1,x_2,x_3,x_4]$ $=\dfrac{f[x_2,x_3,x_4]-f[x_1,x_2,x_3]}{x_4-x_1}$
x_3	f_3		$f[x_2,x_3,x_4]=\dfrac{f[x_3,x_4]-f[x_2,x_3]}{x_4-x_2}$	
		$f[x_3,x_4]=\dfrac{f_4-f_3}{x_4-x_3}$		
x_4	f_4			

Example: Compute f(0.3) for the data,

$$x \quad 0 \quad 1 \quad 3 \quad 4 \quad 7$$
$$f \quad 1 \quad 3 \quad 49 \quad 129 \quad 813$$

using Newton's divided difference formula.

Solution : Divided difference table

x_i	f_i			
0	1			
		2		
1	3		7	
		23		3
3	49		19	
		80		3
4	129		37	
		228		
7	813			

Now Newton's divided difference formula is,

$$f(x)=f[x_0]+(x-x_0)f[x_0,x_1]+(x-x_0)(x-x_1)f[x_0,x_1,x_2]$$

$$+(x-x_0)(x-x_1)(x-x_2)f[x_0,x_1,x_2,x_3]$$

$$f(0.3)=1(0.3-0)2+(0.3)(0.3-1)7+(0.3)(0.3-1)(0.3-3)3$$
$$=1.831$$

Since the given data is for the polynomial $f(x)=3x^3-5x^2+4x+1$ the analytical value is $f(0.3)=1.831$.

The analytical value is matched with the computed value because the given data is for a third degree polynomial and there are five data points available using which one can approximate any data exactly upto fourth degree polynomial.

Properties

1. If f(x) is a polynomial of degree N, then the Nth divided difference of f(x) is a constant.

Proof: Consider the divided difference of x^n

$$\Delta x^n = \frac{(x_1 + h)^n - x^n}{x + h - x} = \frac{n h x^{n-1} + \ldots}{h}$$

= a polynomial of degree (n - 1)

Also since divided difference operator is a linear operator, Δ of any N^{th} degree polynomial is an $(N-1)^{th}$ degree polynomial and second Δ is an (N-2) degree polynomial, so on the Nth divided difference of an Nth degree polynomial is a constant.

2. If $x_0, x_1, x_2 \ldots x_n$ are the (n+1) discrete points then the Nth divided difference is equal to,

$$f[x_0, x_1, x_2 \ldots x_n] = \frac{f_0}{(x_0 - x_1) \cdots (x_0 - x_n)} + \ldots + \frac{f_n}{(x_n - x_0) \cdots (x_n - x_{n-1})}$$

Proof : If $n = 0 \Rightarrow f(x_0) = f(x_0)$ hence the result is true let us assume that the result is valid upto $n = k$,

$$\Rightarrow f[x_0, x_1, \ldots x_k] = \frac{f_0}{(x_0 - x_1) \cdots (x_0 - x_k)} + \ldots + \frac{f_k}{(x_k - x_0) \cdots (x_k - x_{k-1})}$$

Consider the case n = k + 1,

$$\Rightarrow f[x_0, x_1, \ldots x_k, x_{k+1}] = \frac{f[x_1, x_2, \ldots x_{k+1}] - f[x_0, x_1, \ldots x_k]}{(x_{k+1} - x_0)}$$

$$= \frac{1}{(x_{k+1} - x_0)} \left[\frac{f_1}{(x_1 - x_2) \cdots (x_1 - x_{k-1})} + \ldots + \frac{f_{k+1}}{(x_{k+1} - x_1) \cdots (x_{k+1} - x_k)} \right]$$

$$- \frac{1}{(x_{k+1} - x_0)} \left[\frac{f_0}{(x_0 - x_1) \cdots (x_0 - x_k)} + \ldots + \frac{f_k}{(x_k - x_0) \cdots (x_k - x_{k-1})} \right]$$

$$= \frac{f_0}{(x_0 - x_1) \cdots (x_0 - x_{k+1})} + \frac{f_1}{(x_1 - x_2) \cdots (x_1 - x_k)(x_{k+1} - x_0)} \left(\frac{1}{x_1 - x_{k+1}} - \frac{1}{x_1 - x_0} \right) +$$

$$\ldots + \frac{f_{k+1}}{(x_{k+1} - x_0) \cdots (x_{k+1} - x_k)}$$

$$= \frac{f_0}{(x_0 - x_1)\cdots(x_0 - x_{k+1})} + \frac{f_1}{(x_1 - x_0)(x_1 - x_2)\cdots(x_1 - x_{k+1})} +$$

$$\cdots + \frac{f_{k+1}}{(x_{k+1} - x_0)\cdots(x_{k+1} - x_k)}$$

3. Sheppard Zigzag rule:

Consider the divided difference table for the data points $(x_0, f_0), (x_1, f_1), (x_2, f_2)$ and (x_3, f_3)

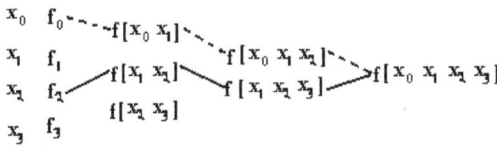

In the difference table the dotted line and the solid line give two differenct paths starting from the function values to the higher divided difference's posssible to the function values. The Sheppard's zigzag rule says the function value at any non-tabulated from the dotted line or from the solid line are same provided the order of xi are taken in the direction of the zigzag line. That is any f(x) through the dotted line can be approximated as

$$f(x) = f_0 + (x - x_0)f[x_0, x_1] + (x - x_0)(x - x_1)f[x_0, x_1, x_2] + (x - x_0)(x - x_1)(x - x_2)f[x_0, x_1, x_2, x_3].$$

Similarly f(x) over the solid line is euivalent to

$$f(x) = f_2 + (x - x_2)f(x_1, x_2) + (x - x_2)(x - x_1)f[x_1, x_2, x_3] + (x - x_2)(x - x_1)(x - x_3)f[x_0, x_1, x_2, x_3].$$

Example: Find f(1.5) from the data points

X	0	0.5	1	2
f(x)	1	1.8987	3.7183	11.3891

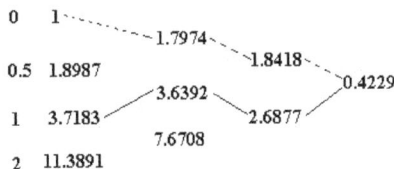

$f(1.5)$ along the dotted line is

$f(1.5) = 1 + 1.5 \times 1.7974 + 1.5\,(1) \times 1.8418 + (1.5)\,(1)\,(0.5) \times 0.4229$

$= 6.770.$

Similarly f(1.5) along the solid line is,

$f(1.5) = 3.7183+(1.5 - 1) \times 3.6392+(1.5 - 1)(1.5 - 0.5) \times 2.6877+(1.5 -1)(1.5 - 0.5)(1.5 - 2) \times 0.4229.$

$= 6.770$

The data is given for $f(x) = x^2 + e^x$ and the analytical value for $f(1.5) = 6.7317$,

The error term is,

$$E(x) = (x - x_0)(x - x_1) \cdots (x - x_n) f[x_0, x_1, \ldots, x_n, x]$$
$$= (x - x_0)(x - x_1) \cdots (x - x_n) \frac{f^{n+1}(\xi)}{(n+1)!},$$

where $\min\{x_0, x_1, \ldots, x_n, x\} < \xi < \max\{x_0, x_1, \ldots, x_n, x\}$.

It is very interesting that the Newton's fundamental interpolation formula gives both the interpolating polynomial as well as error term simultaneously.

Example: Using Newton's divided difference formula, find the value of $f(1.2)$ from the following table:

X	:	0	2	5	7	11
f(x)	:	2.153	3.875	4.279	4.891	5.256

Solution. The divided difference table is

x	f(x)	1st	2nd	3rd	4th
0	2.153				
		-2			
			0.8610		
		-5			
2	3.875		-0.1453		
		-7	-3		
			0.1346	0.0257	
-11		-5			
5	4.279		0.0343		-0.0030
		-9	-2		
			0.3060	-0.0078	
		-6			
7	4.891		-0.0358		
		-4			
			0.0912		
11	5.256				

Here $x = 1.2, x_0 = 0, x_1 = 2, x_2 = 5, x_3 = 7, x_4 = 11$,

$$f[x_0] = 2.153, f[x_0, x_1] = 0.8610, f[x_0, x_1, x_2] = -0.1453,$$

$$f[x_0, x_1, x_2, x_3] = 0.0257, f[x_0, x_1, x_2, x_3, x_4] = -0.0030.$$

The Newton's divided difference formula is,

$$
\begin{aligned}
f(1.2) = {} & f[x_0] + (x - x_0) f[x_0, x_1] + (x - x_0)(x - x_1) f[x_0, x_1, x_2] \\
& + (x - x_0) + (x - x_1)(x - x_2) f[x_0, x_1, x_2, x_3] \\
& + (x - x_0) + (x - x_1)(x - x_2)(x - x_3) f[x_0, x_1, x_2, x_3, x_4] \\
= {} & 2.153 + 1.0332 + 0.1399 + 0.0938 + 0.0635
\end{aligned}
$$

Hence, f(1.2) = 3.4834.

Deductions of other Interpolation Formulae from Newton's Divided Difference Formula

Theoretically, Newton's divided difference formula is very powerful because from this formula several interpolation formulae can be derived.

Newton's Forward Difference Interpolation Formula

Let the arguments be equispaced, i.e. $x_i = x_0 + ih, i = 0, 1, \ldots, n$. Then the kth order divided difference is,

$$f[x_0, x_1, \ldots, x_k] = \frac{\Delta^k f(x_0)}{k! h^k},$$

for $k = 1, 2, \ldots, n$.

Using this assumption, the Newton's general interpolation formula reduces to:

$$
\begin{aligned}
\phi(x) = {} & f(x_0) + (x - x_0) \frac{\Delta f(x_0)}{1! h} + (x - x_0)(x - x_1) \frac{\Delta^2 f(x_0)}{2! h^2} \\
& + (x - x_0)(x - x_1)(x - x_2) \frac{\Delta^3 f(x_0)}{3! h^3} + \ldots \\
& + (x - x_0)(x - x_1) \ldots (x - x_{n-1}) \frac{\Delta^n f(x_0)}{n! h^n} + (x - x_0)(x - x_1) \ldots (x - x_n) \frac{\Delta^{n+1} f(\xi)}{(n+1)! h^{n+1}}.
\end{aligned}
$$

The last term is the error term.

To convert it to usual form of Newton's forward difference formula, let $u = \dfrac{x - x_0}{h}$, i.e. $x = x_0 + uh$. Since $x_i = x_0 + ih, i = 0, 1, 2, \ldots, n$.

Therefore, $x - x_i = (u - i)h$.

Then

$$\phi(x) = f(x_o) + u\Delta f(x_o) + \frac{u(u-1)}{2!}\Delta^2 f(x_o) + \cdots$$

$$+ \frac{u(u-1)(u-2)\cdots(u-n+1)}{n!}\Delta^n f(x_o)$$

$$+ u(u-1)(u-2)\cdots(u-n)\frac{f^{n+1}(\xi)}{(n+1)!}.$$

The value of ξ lies between $\min\{x, x_o, x_1, \ldots x_n\}$ and $\max\{x, x_o, x_1, \ldots x_n\}$.

This is the well known Newton's forward difference interpolation formula including error term.

Newton's Backward Difference Interpolation Formula

In this case also, we assumed that $x_i = x_o + ih, i = 0, 1, \ldots, n$.

Let

$$\phi(x) = f(x_n) + (x - x_n)f[x_n, x_{n-1}] + (x - x_n)(x - x_{n-1})f[x_n, x_{n-1}, x_{n-2}]$$

$$+ \cdots + (x - x_n)(x - x_{n-1})\cdots(x - x_1)f[x_n, x_{n-1}, \ldots, x_1, x_o]$$

$$+ E(x),$$

Where

$$E(x) = (x - x_n)(x - x_{n-1})\cdots(x - x_1)(x - x_o)f[x, x_n, x_{n-1}, \ldots, x_1, x_o].$$

From the arguments $x_n, x_{n-1}, \ldots, x_{n-k}$, the relation $f[x_o, x_1, \ldots, x_n] = \frac{1}{h^n . n!}\Delta^n y_o$. becomes,

$$f[x_n, x_{n-1}, \ldots, x_{n-k}] = \frac{\Delta^k f(x_{n-k})}{k! h^k}.$$

Again,

$$\frac{\Delta^k f(x_{n-k})}{k! h^k} = \frac{\nabla^k f(x_n)}{k! h^k}.$$

Thus,

$$f[x_n, x_{n-1}, \ldots, x_{n-k}] = \frac{\nabla^k f(x_n)}{k! h^k}.$$

With this notation $\phi(x) = f(x_n) + (x - x_n) f[x_n, x_{n-1}] + (x - x_n)(x - x_{n-1}) f[x_n, x_{n-1}, x_{n-2}]$
$$+ \cdots + (x - x_n)(x - x_{n-1}) \cdots (x - x_1) f[x_n, x_{n-1}, \ldots, x_1, x_0]$$
$$+ E(x),$$

reduces to

$$\phi(x) = f(x_n) + (x - x_n)\frac{\nabla f(x_n)}{1!h} + (x - x_n)(x - x_n)\frac{\nabla f(x_n)}{2!h} + \cdots$$
$$+ (x - x_n)(x - x_{n-1}) \quad (x - x_1)(x - x_0)\frac{\nabla f(x)}{n\ h} + E(x),$$

where

$$E(x) = (x - x_n)(x - x_{n-1}) \cdots (x - x_1)(x - x_0)\frac{\nabla^{n+1} f(\xi)}{(n+1)!h^{n+1}},$$

where $\min\{x, x_0, x_1, \ldots x_n\} < \xi < \max\{x, x_0, x_1, \ldots x_n\}$.

This is the Newton's backward difference interpolation formula with error term E(x).

The problems in this section are solved using Newton's divided difference formula and Lagrange's formula. Since By Sheperd's Zig-Zag rule any aritrary path from function values to its highest divided difference to compute the value of f(x) in all these examples first fuction value and its higher divided differences are been used to compute f(x).

Find f(2) for the data f(0) = 1, f(1) = 3 and f(3) = 55.

x	0	1	3
f	1	3	55

Solution :

By Newton's divided difference formula :

Divided difference table

x_i	f_i		
0	1		
		2	
1	3		8
		26	
3	55		

Now Newton's divided difference formula is

$$f(x) = f[x_0] + (x - x_0) f[x_0, x_1] + (x - x_0)(x - x_1) f[x_0, x_1, x_2]$$
$$f(2) = 1 + (2 - 0)2 + (2 - 0)(2 - 1)8$$
$$= 21$$

Find f(3) for

x	0	1	2	4	5	6
f	1	14	15	5	6	19

Solution :

By Newton's divided difference formula:

Divided difference table

x_i	f_i					
0	1					
		13				
1	14		-6			
		1		1		
2	15		-2		0	
		-5		1		0
4	5		2		0	
		1		1		
5	6		6			
		13				
6	19					

Now Newton's divided difference formula is,

$$f(x) = f[x_0] + (x - x_0) f[x_0, x_1] + (x - x_0)(x - x_1) f[x_0, x_1, x_2] + (x - x_0)(x - x_1)(x - x_2) f[x_0, x_1, x_2, x_3]$$
$$+ (x - x_0)(x - x_1)(x - x_2)(x - x_3) f[x_0, x_1, x_2, x_3, x_4]$$
$$+ (x - x_0)(x - x_1)(x - x_2)(x - x_3)(x - x_4) f[x_0, x_1, x_2, x_3, x_4, x_5]$$

$$f(3) = 1 + (3 - 0) 13 + (3 - 0)(3 - 1) - 6 + (3 - 0)(3 - 1)(3 - 2) 1$$
$$= 10$$

Find f(0.25) for

x	0.1	0.2	0.3	0.4	0.5
f	9.9833	4.9667	3.2836	2.4339	1.9177

Solution:

By Newton's divided difference formula:

Divided difference table

x_i	f_i				
0.1	9.9833				
		-50.166			
0.2	4.9667		166.675		

		-16.83		-416.68	
0.3	3.2836		41.67		833.42
		-8.497		-83.32	
0.4	2.4339		16.675		
		-5.162			
0.5	1.9177				

Now Newton's divided difference formula is,

$$f(x) = f[x_0] + (x - x_0) f[x_0, x_1] + (x - x_0)(x - x_1) f[x_0, x_1, x_2] + (x - x_0)(x - x_1)(x - x_2) f[x_0, x_1, x_2, x_3]$$
$$+ (x - x_0)(x - x_1)(x - x_2)(x - x_3) f[x_0, x_1, x_2, x_3, x_4]$$

$$f(3) = 9.9833 + (0.25 - 0.1) - 50.166 + (0.25 - 0.2)(0.25 - 0.3)166.675 +$$
$$(0.25 - 0.1)(0.25 - 0.2)(0.25 - 0.3) - 416.68 + (0.25 - 0.1)(0.25 - 0.2)(0.25 - 0.3)(0.25 - 0.4)833.42$$
$$= 3.912$$

Example:

Find the polynomial P_2 for the function $y = \sqrt{x}$ that interpolates the points (1,1), (4,2), and (9,3) using Newton's divided difference formula.

Applying Newton's divided difference formula from above and we get that:

$$P_2(x) = f(x_0) + (x - x_0) f[x_0, x_1] + (x - x_0)(x - x_1) f[x_0, x_1, x_2]$$
$$P_2(x) = 1 + (x - 1)\frac{f(x_1) - f(x_0)}{x_1 - x_0} + (x - 1)(x - 4)\frac{f[x_1, x_2] - f[x_0, x_1]}{x_2 - xo}$$
$$P_2(x) = 1 + (x - 1)\frac{2 - 1}{4 - 1} + (x - 1)(x - 4)\frac{\frac{3 - 2}{9 - 4} - \frac{2 - 1}{4 - 1}}{9 - 1}$$
$$P_2(x) = 1 + \frac{1}{3}(x - 1) - \frac{1}{60}(x - 1)(x - 4)$$

The graph of $y = P_2(x)$ is given below:

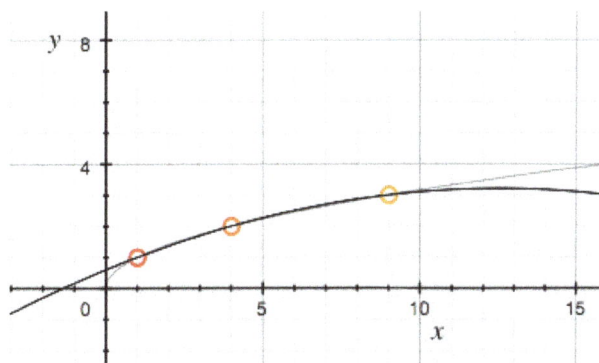

Example:

Find the polynomial P_3 for the function $y = \sin x$ that interpolates the points $(1, \sin 1)$, $\left(\dfrac{\pi}{2}, 1\right)$, $(2, \sin 2)$ and $(\pi, 0)$ using Newton's divided difference formula.

Applying Newton's divided different formula from above and we get that:

$$P_3(x) = f(x_0) + (x - x_0)f[x_0, x_1] + (x - x_0)(x - x_1)f[x_0, x_1, x_2] + (x - x_0)(x - x_1)(x - x_2)f[x_0, x_1, x_2, x_3]$$

$$P_3(x) = \sin 1 + (x-1)\frac{1-\sin 1}{\frac{\pi}{2}-1} + (x-1)\left(x-\frac{\pi}{2}\right)\frac{\frac{\sin 2-1}{2-\frac{\pi}{2}} - \frac{1-\sin 1}{\frac{\pi}{2}-1}}{2-1} + (x-1)\left(x-\frac{\pi}{2}\right)(x-2)\frac{f[x_1,x_2,x_3]-f[x_0,x_1,x_2]}{x_3-x_0}$$

$$P_3(x) = \sin 1 + (x-1)\frac{1-\sin 1}{\frac{\pi}{2}-1} + (x-1)\left(x-\frac{\pi}{2}\right)\frac{\frac{\sin 2-1}{2-\frac{\pi}{2}} - \frac{1-\sin 1}{\frac{\pi}{2}-1}}{2-1} + (x-1)\left(x-\frac{\pi}{2}\right)(x-2)\frac{\frac{f[x_2,x_3]-f[x_1,x_2]}{x_3-x_1} - \frac{f[x_1,x_2]-f[x_0,x_1]}{x_2-x_0}}{x_3-x_0}$$

$$P_3(x) = \sin 1 + (x-1)\frac{1-\sin 1}{\frac{\pi}{2}-1} + (x-1)\left(x-\frac{\pi}{2}\right)\frac{\frac{\sin 2-1}{2-\frac{\pi}{2}} - \frac{1-\sin 1}{\frac{\pi}{2}-1}}{2-1} + (x-1)\left(x-\frac{\pi}{2}\right)(x-2)\frac{\left(\frac{\frac{0-\sin 2}{\pi-2} - \frac{\sin 2-1}{2-\frac{\pi}{2}}}{\pi-\frac{\pi}{2}}\right) - \left(\frac{\frac{\sin 2-1}{2-\frac{\pi}{2}} - \frac{1-\sin 1}{\frac{\pi}{2}-1}}{2-1}\right)}{\pi-1}$$

The graph of $y = P_3(x)$ is given below.

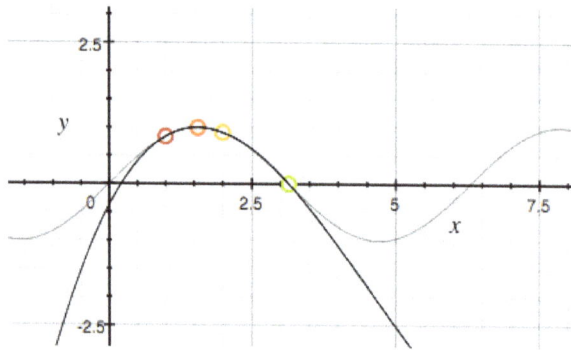

Lagrange Polynomial

The Lagrange interpolation formula is a way to find a polynomial which takes on certain values at arbitrary points.

Let us suppose that the given data points $(x_i, y_i), i = 0, 1, 2 \ldots n$ is coming from a function $f(x)$. Let us assume that this function $y = f(x)$ takes the values $y_0, y_1 \ldots \ldots y_n$ at $x_0, x_1, \ldots \ldots x_n$. Since there are $(n+1)$ data points (x_i, y_i), we can represent the function $f(x)$ by a polynomial of degree n,

$$\therefore f(x) = C_n x^n + C_{n-1} x^{n-1} + \ldots + C_1 x + C_0$$

As we have assumed that $f(x_i) = y_i, i = 0,1,2\ldots\ldots n$ i.e. the function $f(x)$ passes through (x_i, y_i), above equation can be rewritten as:

$$y = f(x) = a_0(x-x_1)(x-x_2)\ldots(x-x_n) + a_1(x-x_0)(x-x_2)\ldots(x-x_n) +$$

$$a_2(x-x_0)(x-x_1)(x-x_3)\ldots(x-x_n) + \ldots\ldots + a_n(x-x_0)\ldots(x-x_{n-1})$$

$$But, y_i = f(x_i) \quad i = 0,1,\ldots..n$$

Using above equation for i=0, in
$a_2(x-x_0)(x-x_1)(x-x_3)\ldots(x-x_n) + \ldots\ldots + a_n(x-x_0)\ldots(x-x_{n-1})$ we get,

$$y_1 = f(x_1) = a_1(x_0 - x_1)\ldots(x_1 - x_n)$$

$$\therefore a_1 = \frac{y_0}{(x_1 - x_0)(x_1 - x_2)\ldots(x_1 - x_n)}$$

For $i=1$ we get,

$$y_1 = f(x_1) = a_1(x_0 - x_1)\ldots(x_1 - x_n)$$

$$\therefore a_1 = \frac{y_0}{(x_1 - x_0)(x_1 - x_2)\ldots(x_1 - x_n)}$$

Similarly for $i = 2\ldots\ldots n-1$, we get,

$$a_i = \frac{y_i}{(x_i - x_0)(x_i - x_1)\ldots(x_i - x_{i-1})(x_i - x_{i+1})\ldots(x_i - x_n)}$$

and for $i = n$, we get,

$$a_n = \frac{y_n}{(x_n - x_0)\ldots(x_n - x_{n-1})}$$

Using above equations in $a_2(x-x_0)(x-x_1)(x-x_3)\ldots(x-x_n) + \ldots\ldots + a_n(x-x_0)\ldots(x-x_{n-1})$ we get,

$$y = f(x) = \frac{(x-x_1)(x-x_2)\ldots(x-x_n)}{(x_0-x_1)(x_0-x_2)\ldots(x_0-x_n)} y_0 + \frac{(x-x_0)(x-x_2)\ldots(x-x_n)}{(x_1-x_0)(x_1-x_2)\ldots(x_1-x_n)} y_1 + \ldots\ldots$$

$$+ \frac{(x-x_0)(x-x_1)\ldots(x-x_{i-1})(x-x_{i+1})\ldots(x-x_n)}{(x_i-x_0)(x_i-x_1)\ldots(x_i-x_{i-1})(x_i-x_{i+1})\ldots(x-x_n)} y_i + \ldots + \frac{(x-x_0)(x-x_1)\ldots(x-x_{n-1})}{(x_n-x_0)(x_n-x_1)\ldots(x_n-x_{n-1})} y_n$$

Above equation can be rewritten in a compact form as:

$$y = f(x) = L_0(x)y_0 + L_1(x)y_1 + \ldots\ldots + L_n(x)y_n$$

$$= \sum_{i=0}^{n} L_i(x)y_i$$

$$= \sum_{i=0}^{n} L_i(x)f(x_i)$$

where

$$L_i(x) = \frac{(x-x_0)(x-x_1)\ldots(x-x_{i-1})(x-x_{i+1})\ldots(x-x\ldots\ldots)}{(x_i-x_0)(x_i-x_1)\ldots(x_i-x_{i-1})(x_i-x_{i+1})\ldots(x_i-x_n)}$$

It can be easily noted that,

$$L_i(x_j) = \begin{cases} 1 \text{ if } i = j \\ 1 \text{ if } i \neq j \end{cases}$$

Let us introduce the product notation as:

$$\prod(x) = \prod_{i=0}^{n}(x-x_i) = (x-x_0)(x-x_1)\ldots(x-x_n)$$

$$\therefore L_k(x) = \frac{\prod\limits_{i=0,i\neq k}^{n}(x-x_i)}{\prod\limits_{i=0,i\neq k}^{n}(x_k-x_i)}$$

Therefore, Lagrange interpolation polynomial of degree n can be written as,

$$y = f(x) = \sum_{k=0}^{n} L_k(x)y_k$$

Example:

Given the following data table, construct the lagrange interpolation polynomial $f(x)$, to fit the data and find $f(1.25)$:

i	0	1	2	3
x_i	0	1	2	3
$y_i = f(x_i)$	1	2.25	3.75	4.25

Solution:

Here $n = 3$.

$\therefore$ Lagrange interpolation polynomial is given by,

$$y = f(x) = \sum_{i=0}^{3} L_i(x)y_i$$

$$L(x) \quad \frac{\displaystyle\prod_{i=0,i\neq 0}(x \quad x)}{\displaystyle\prod_{i=0,i\neq 0}(x \quad x)}$$

$$= \frac{(x-x_1)(x-x_2)(x-x_3)}{(x_0-x_1)(x_0-x_2)(x_0-x_3)}$$

$$= \frac{(x-1)(x-2)(x-3)}{(0-1)(0-2)(0-3)}$$

$$= \frac{x^3 - 6x^2 + 11x - 6}{-6}$$

$$L_1(x) = \frac{\displaystyle\prod_{i=0,i\neq 1}^{3}(x-x_i)}{\displaystyle\prod_{i=0,i\neq 1}^{3}(x_1-x_i)}$$

$$= \frac{(x-x_0)(x-x_2)(x-x_3)}{(x_1-x_0)(x_1-x_2)(x_1-x_3)}$$

$$= \frac{(x-0)(x-2)(x-3)}{(1-0)(1-2)(1-3)}$$

$$= \frac{x^3 - 5x^2 + 6x}{2}$$

$$L_2(x) = \frac{\displaystyle\prod_{i=0,i\neq 2}^{3}(x-x_i)}{\displaystyle\prod_{i=0,i\neq 2}^{3}(x_2-x_i)}$$

$$= \frac{(x-x_0)(x-x_1)(x-x_3)}{(x_2-x_0)(x_2-x_1)(x_2-x_3)}$$

$$= \frac{(x-0)(x-1)(x-3)}{(2-0)(2-1)(2-3)}$$

$$= \frac{x^3 - 4x^2 + 3x}{-2}$$

$$L_3(x) = \frac{\displaystyle\prod_{i=0,i\neq 3}^{3} (x-x_i)}{\displaystyle\prod_{i=0,i\neq 3}^{3} (x_3 - x_i)}$$

$$= \frac{(x-x_0)(x-x_1)(x-x_2)}{(x_3-x_0)(x_3-x_1)(x_3-x_2)}$$

$$= \frac{(x-0)(x-1)(x-2)}{(3-0)(3-1)(3-2)}$$

$$= \frac{x^3 - 3x^2 + 2x}{6}$$

$$\therefore \quad f(1.25) = \sum_{i=0}^{3} L_i(1.25)y_i$$

$$= L_0(1.25)y_0 + L_1(1.25)y_1 + L_2(1.25)y_2 + L_3(1.25)y_3$$

$$= (-0.546875).1 + (0.8203125)2.25 + (0.2734375)3.75 + (-0.0390625)4.25$$

$$= 2.650390625$$

Example:

Given the following data table, construct the Lagrange interpolation polynomial f(x), to fit the data and find $f(1998)$:

i	0	1	2	3	4	5
x_i	1980	1985	1990	1995	2000	2005
$y_i = f(x_i)$	440	510	525	571	500	600

Solution:

Here $n = 6$, $\quad x_k = 1998$

$\therefore$ Lagrange interpolation polynomial is given by,

$$y = f(x) = \sum_{i=0}^{5} L_i(x) y_i; \quad y_k = f(x_k) = \sum_{i=0}^{5} L_i(x) y_i$$

$$L_0(x) = \frac{\prod\limits_{i=0, i \neq 0}^{5} (x - x_i)}{\prod\limits_{i=0, i \neq 0}^{5} (x_0 - x_i)} = \frac{(x - x_1)(x - x_2)(x - x_3)(x - x_4)(x - x_5)}{(x_0 - x_1)(x_0 - x_2)(x_0 - x_3)(x_0 - x_4)(x_0 - x_5)}$$

$$= \frac{(x - 1985)(x - 1990)(x - 1995)(x - 2000)(x - 2005)}{(1980 - 1985)(1980 - 1990)(1980 - 1995)(1980 - 2000)(1980 - 2005)}$$

$$L_0(x_k) = L_0(1998) = \frac{(1998 - 1985)(1998 - 1990)(1998 - 1995)(1998 - 2000)(1998 - 2005)}{(-5)(-10)(-15)(-20)(-25)}$$

$$= \frac{13.8.3.(-2).(-7)}{-(375000)}$$

$$= -\frac{4368}{375000} = -0.011648$$

$$L_1(x_k) = \frac{\prod\limits_{i=0, i \neq 1}^{5} (x_k - x_i)}{\prod\limits_{i=0, i \neq 1}^{5} (x_1 - x_i)} = \frac{(x_k - x_0)(x_k - x_2)(x_k - x_3)(x_k - x_4)(x_k - x_5)}{(x_1 - x_0)(x_1 - x_2)(x_1 - x_3)(x_1 - x_4)(x_1 - x_5)}$$

$$= \frac{(1998 - 1980)(1998 - 1990)(1998 - 1995)(1998 - 2000)(1998 - 2005)}{(1985 - 1980)(1985 - 1990)(1985 - 1995)(1985 - 2000)(1985 - 2005)}$$

$$= \frac{18.8.3.(-2).(-7)}{5(-5)(-10)(-15)(-20)} = 0.08064$$

$$L_3(x_k) = \frac{\prod\limits_{i=0, i \neq 3}^{5} (x_k - x_i)}{\prod\limits_{i=0, i \neq 3}^{5} (x_3 - x_i)} = \frac{(1998 - 1980)(1998 - 1985)(1998 - 1995)(1998 - 2000)(1998 - 2005)}{(1990 - 1980)(1990 - 1985)(1990 - 1995)(1990 - 2000)(1990 - 2005)}$$

$$= \frac{18.13.3.(-2)(-7)}{10.5.(-5)(-10).(-15)}$$

$$= -0.26208$$

$$L_3(x_k) = \frac{\prod\limits_{i=0, i \neq 3}^{5} (x_k - x_i)}{\prod\limits_{i=0, i \neq 3}^{5} (x_3 - x_i)} = \frac{(1998 - 1980)(1998 - 1985)(1998 - 1990)(1998 - 2000)(1998 - 2005)}{(1995 - 1980)(1995 - 1985)(1995 - 1990)(1995 - 2000)(1995 - 2005)}$$

$$= \frac{18.13.8.(-2)(-7)}{15.10.5(-5)(-10)}$$

$$= 0.69888$$

$$L_4(x_k) = \frac{\prod_{i=0,i\neq 4}^{5}(x_k - x_i)}{\prod_{i=0,i\neq 4}^{5}(x_4 - x_i)} = \frac{(1998-1980)(1998-1985)(1998-1990)(1998-1995)(1998-2005)}{(2000-1980)(2000-1985)(2000-1990)(2000-1995)(2000-2005)}$$

$$= \frac{18.13.8.3.(-7)}{20.15.10.5(-5)} = 0.52416$$

$$L_5(x_k) = \frac{\prod_{i=0,i\neq 5}^{5}(x_k - x_i)}{\prod_{i=0,i\neq 5}^{5}(x_5 - x_i)} = \frac{(1998-1980(1998-1985)(1998-1990)(1998-1995)(1998-2000)}{(2005-1980)(2005-1985)(2005-1990)(2005-1995)(2005-2000)}$$

$$= \frac{18.13.8.3.(-2)}{25.20.15.10.5} = -0.029952$$

$$\therefore \ f(1998) = \sum_{i=0}^{5} L_i(1998)y_i$$

$$= -0.011648 \times 440 + 0.08064 \times 510 + (-0.26208) \times 525 + 0.69888 \times 571$$

$$+ 0.52416 \times 500 + (-0.029952) \times 600$$

$$= 541.578560$$

Given a set of data points (x_i, y_i), $i = 1, \ldots n$. Suppose we are interested in evaluating $f(x)$ at some intermediate point x to a desired level of accuracy. Directly using the entire data set of size n may not only be computationally economical but may also turn out to be redundant. Naturally one would like to use an interpolating polynomial of optimal degree. Since this is not known apriori, one may start with $P_0(x)$ and if it was enough then move onto $P_1(x)$ and so on i.e. slowly increase the no. of the interpolating points (or) data points $x_0, x_1 .. x_k$ so that $P_{k-1}(x)$ will be close to $f(x)$. In this context the biggest disadvantage with Lagrange Interpolation is that we cannot use the work that has already been done i.e. we cannot make use of $P_{k-1}(x)$ while evaluating $P_k(x)$. With the addition of each new data point, calculations have to be repeated. Newton Interpolation polynomial overcomes this drawback.

Barycentric form

Using

$$\ell(x) = (x - x_0)(x - x_1) \cdots (x - x_k)$$

$$\ell'(x_j) = \frac{d\ell(x)}{dx}\Big|_{x=x_j} = \prod_{i=0, i \neq j}^{k} (x_j - x_i)$$

we can rewrite the Lagrange basis polynomials as,

$$\ell_j(x) = \frac{\ell(x)}{\ell'(x_j)(x - x_j)}$$

or, by defining the *barycentric weights*,

$$w_j = \frac{1}{\ell'(x_j)}$$

we can simply write,

$$\ell_j(x) = \ell(x)\frac{w_j}{x - x_j}$$

which is commonly referred to as the *first form* of the barycentric interpolation formula.

The advantage of this representation is that the interpolation polynomial may now be evaluated as,

$$L(x) = \ell(x)\sum_{j=0}^{k}\frac{w_j}{x - x_j}y_j$$

which, if the weights w_j have been pre-computed, requires only $\mathcal{O}(n)$ operations (evaluating $\ell(x)$ and the weights $w_j/(x - x_j)$) as opposed to $\mathcal{O}(n^2)$ for evaluating the Lagrange basis polynomials $\ell_j(x)$ individually.

The barycentric interpolation formula can also easily be updated to incorporate a new node x_{k+1} by dividing each of the w_j, $j = 0 \ldots k$ by $(x_j - x_{k+1})$ and constructing the new w_{k+1} as above.

We can further simplify the first form by first considering the barycentric interpolation of the constant function $g(x) \equiv 1$:

$$g(x) = \ell(x)\sum_{j=0}^{k}\frac{w_j}{x - x_j}.$$

Dividing $L(x)$ by $g(x)$ does not modify the interpolation, yet yields,

$$L(x) = \frac{\displaystyle\sum_{j=0}^{k}\frac{w_j}{x - x_j}y_j}{\displaystyle\sum_{j=0}^{k}\frac{w_j}{x - x_j}}$$

which is referred to as the *second form* or *true form* of the barycentric interpolation formula. This second form has the advantage that $\ell(x)$ need not be evaluated for each evaluation of $L(x)$.

Remainder in Lagrange Interpolation Formula

When interpolating a given function f by a polynomial of degree n at the nodes $x_0,...,x_n$ we get the remainder $R(x) = f(x) - L(x)$ which can be expressed as,

$$R(x) = f[x_0,...,x_n,x]\ell(x) = \ell(x)\frac{f^{n+1}(\xi)}{(n+1)!}, \qquad x_0 < \xi < x_n,$$

where $f[x_0,...,x_n,x]$ is the notation for divided differences. Alternatively, the remainder can be expressed as a contour integral in complex domain as,

$$R(z) = \frac{\ell(z)}{2\pi i}\int_C \frac{f(t)}{(t-z)(t-z_0)\cdots(t-z_n)}dt = \frac{\ell(z)}{2\pi i}\int_C \frac{f(t)}{(t-z)\ell(t)}dt.$$

The remainder can be bound as,

$$|R(x)| \le \frac{(x_n - x_0)^{n+1}}{(n+1)!}\max_{x_0 \le \xi \le x_n} |f^{(n+1)}(\xi)|.$$

Derivatives

The dth derivatives of the lagrange polynomial can be written as,

$$L^{(d)}(x) := \sum_{j=0}^{k} y_j \ell_j^{(d)}(x).$$

For the first derivative, the coefficients are given by,

$$\ell_j^{(1)}(x) := \sum_{i=0,i\neq j}^{k}\left[\frac{1}{x_j - x_i}\prod_{m=0,m\neq(i,j)}^{k}\frac{x - x_m}{x_j - x_m}\right]$$

and for the second derivative,

$$\ell_j^{(2)}(x) := \sum_{i=0,i\neq j}^{k}\frac{1}{x_j - x_i}\left[\sum_{m=0,m\neq(i,j)}^{k}\left(\frac{1}{x_j - x_m}\prod_{l=0,l\neq(i,j,m)}^{k}\frac{x - x_l}{x_j - x_l}\right)\right].$$

Through recursion, one can compute formulas for higher derivatives.

Theorem

Given n distinct real values $x_1, x_2, ..., x_n$ and n real values $y_1, y_2, ..., y_n$ (not necessarily distinct), there is a unique polynomial P with real coefficients satisfying $P(x_i) = y_i$ for $i \in \{1, 2, \ldots, n\}$, such that $\deg(P) < n$.

This theorem can be viewed as a generalization of the well-known fact that two points uniquely determine a straight line, three points uniquely determine the graph of a quadratic polynomial, four points uniquely determine the graph of a cubic polynomial, and so on. (Two caveats: (1) the points are required to have different x-coordinates, and (2) the "quadratic polynomial" might actually be a linear or constant polynomial, the "cubic polynomial" might actually be a quadratic, linear, or constant polynomial, and so on.)

Proof

First, a proof that the polynomial P is unique: Suppose Q and R are polynomials with the above properties. Then $Q - R$ vanishes on $x_1, x_2, \ldots, x_n$, but its degree is less than n. A nonzero polynomial of degree $< n$ cannot have n roots, so $Q - R$ must be the zero polynomial, i.e. $Q = R$.

Now, to show that P exists, let

$$P_1(x) \frac{(x - x_2)(x - x_3)(\ldots)(x - x_n)}{(x_1 - x_2)(x_1 - x_3)(\ldots)(x_1 - x_n)}.$$

Then $P_1(x) = 1$ and $P_1(x_2) = P_1(x_3) = \cdots = P_1(x_n) = 0$.

Similarly construct polynomials $P_2, P_2, \ldots, P_n$ such that $P_j(x_j) = 1$ and $P_j(x_i) = 0$ for all $i \neq j$. One way to write $P_j(x)$ is,

$$P_j(x) = \frac{f(x)}{(x - x_j)f'(x_j)},$$

where $f(x) = (x - x_1)(x - x_2)(\cdots)(x - x_n)$.

Then, $P_j(x) = \sum y_i P_i(x)$ is a polynomial with real coefficients satisfying $P(x_i) = y_i$ for all $i \in \{1, 2, \ldots n\}$. It is a sum of polynomials of degree $n - 1$, so its degree is $< n$.

Also note that the theorem still holds, with the same proof as above, if any field is substituted for the real numbers (e.g. the rational numbers, the complex numbers, the integers mod p, etc.).

Cubic Spline Interpolation

A cubic spline is a spline constructed of piecewise third-order polynomials which pass through a set of m control points. The second derivative of each polynomial is commonly set to zero at the endpoints, since this provides a boundary condition that completes the system of m-2 equations. This produces a so-called "natural" cubic spline and leads to a simple tridiagonal system which can be solved easily to give the coefficients of the polynomials.

Process

The essential idea is to fit a piecewise function of the form,

$$S(x) = \begin{cases} s_1(x) & \text{if} \quad x_1 \le x < x_2 \\ s_2(x) & \text{if} \quad x_2 \le x < x_3 \\ \quad \vdots \\ s_{n-1}(x) & \text{if} \quad x_{n-1} \le x < x_n \end{cases}$$

where s_i is a third degree polynomial defined by,

$$S_i(x) = a_i(x - x_i)^3 + b_i(x - x_i)^2 + c_i(x - x_i) + d_i$$

for $i = 1, 2, ..., n-1$.

The first and second derivatives of these $n - 1$ equations are fundamental to this process, and they are,

$$S_i'(x) = 3a_i(x - x_i)^2 + 2b_i(x - x_i)$$
$$S_i''(x) = 6a_i(x - x_i) + 2b_i$$

for $i = 1, 2, ..., n-1$.

Four Properties of Cubic Splines

Our spline will need to conform to the following stipulations.

- The piecewise function $S(x)$ will interpolate all data points.
- $S(x)$ will be continuous on the interval $[x_1, x_n]$
- $S'(x)$ will be continuous on the interval $[x_1, x_n]$
- $S''(x)$ will be continuous on the interval $[x_1, x_n]$

Since the piecewiece function $S(x)$ will interpolate all of the data points, we can conclude that,

$$S(x_i) = y_i$$

for $i = 1, 2, ..., n-1$. Since $x_i \in [x_i, x_{i+1}], S(x_i) = s_i(x_i)$ and we can use equation $S_i(x) = a_i(x - x_i)^3 + b_i(x - x_i)^2 + c_i(x - x_i) + d_i$ to produce,

$$y_i = s_i(x_i)$$

$$y_i = a_i(x_i - x_i)^3 + b_i(x_i - x_i)^2 + c_i(x_i - x_i) + d_i$$

$$y_i = d_i$$

for each $i = 1, 2, \ldots, n-1$. Since the curve $S(x)$ must be continuous across its entire interval, it can be concluded that each sub-function must join at the data points, so

$$s_i(x_i) = s_{i-1}(x_i)$$

for $i = 2, 3, \ldots, n$.

From equation ($S_i(x) = a_i(x - x_i)^3 + b_i(x - x_i)^2 + c_i(x - x_i) + d_i$),

$$s_i(x_i) = d_i$$

and

$$s_{i-1}(x_i) = a_{i-1}(x_i - x_{i-1})^3 + b_{i-1}(x_i - x_{i-1})^2 + c_{i-1}(x_i - x_{i-1}) + d_{i-1}$$

so

$$d_i = a_{i-1}(x_i - x_{i-1})^3 + b_{i-1}(x_i - x_{i-1})^2 + c_{i-1}(x_i - x_{i-1}) + d_{i-1}$$

for $i = 2, 3, \ldots, n-1$. Letting $h = x_i - x_{i-1}$ in equation mentioned above the last one, we have

$$d_i = a_{i-1}h^3 + b_{i-1}h^2 + c_{i-1}h + d_{i-1}$$

for $i = 2, 3, \ldots, n-1$. Also, to make the curve smooth across the interval, the derivatives must be equal at the data points; that is,

$$s_i'(x_i) = s_{i-1}'(x_i)$$

However, by equation ($S_i'(x) = 3a_i(x - x_i)^2 + 2b_i(x - x_i)$),

$$s_i'(x_i) = c_i$$

and

$$s_{i-1}'(x_i) = 3a_{i-1}(x_i - x_{i-1})^2 + 2b_{i-1}(x_i - x_{i-1}) + c_{i-1}$$

$$c_i = 3a_{i-1}(x_i - x_{i-1})^2 + 2b_{i-1}(x_i - x_{i-1}) + c_{i-1}.$$

Again, letting $h = x_i - x_{i-1}$, we arrive at,

$$c_i = 3a_{i-1}h^2 + 2b_{i-1}h + c_{i-1}$$

for $i = 2, 3, \ldots, n-1$.

From equation ($S_i''(x) = 6a_i(x - x_i) + 2b_i$), $s_i''(x) = 6a_i(x - x_i) + 2b_i$, so

$$s_i''(x) = 6a_i(x - x_i) + 2b_i$$
$$s_i''(x_i) = 6a_i(x_i - x_i) + 2b_i$$
$$s_i''(x_i) = 2b_i$$

for $i = 2, 3, \ldots, n-2$.

Lastly, since $s_i''(x)$ has to be continuous across the interval, $s_i''(x_i) = s_{i+1}''(x_i)$ for $i = 1, 2, 3, \ldots, n-1$. This and equation above lead us to the equation,

$$s_i''(x_{i+1}) = 6a_i(x_{i+1} - x_i) + 2b_i$$
$$s_{i+1}''(x_{i+1}) = 6a_i(x_{i+1} - x_i) + 2b_i$$

and, letting $h = x_{i-1} - x_i$ i and using the conclusion from equations mentioned above the last one and equation above,

$$s_{i+1}''(x_{i+1}) = 6a_i(x_{i+1} - x_i) + 2b_i$$
$$2b_{i+1} = 6a_ih + 2b_i$$

These equations can be much simplified by substituting M_i for $s_i''(x_i)$ and expressing the above equations in terms of M_i and y_i. This makes the determination of the weights a_i, b_i, c_i, and d_i a much easier task. Each b_i can be represented by,

$$s_i''(x_i) = 2b_i$$
$$M_i = 2b_i$$
$$b_i = \frac{M_i}{2}$$

and d_i has already been determined to be,

$$d_i = y_i.$$

Similarly, using equation a_i can be re-written as,

$$2b_{i+1} = 6a_ih + 2b_i$$
$$6a_{i+1}h = 2b_{i+1} - 2b_i$$

$$a_i = \frac{2b_{i+1} - 2b_i}{6h}$$

$$a_i = \frac{2(\frac{M_{i+1}}{2}) - 2(\frac{M_i}{2})}{6h}$$

$$a_i = \frac{M_{i+1} - M_i}{6h}$$

and c_i can be re-written as,

$$d_{i+1} = a_i h^3 + b_i h^2 + c_i h + d_i$$

$$c_i h = -a_i h^3 - b_i h^2 - d_i + d_{i+1}$$

$$c_i = \frac{-a_i h^3 - b_i h^2 - d_i + d_{i+1}}{h}$$

$$c_i = \frac{-a_i h^3 - b_i h^2}{h} + \frac{-d_i - d_{i+1}}{h}$$

$$c_i = (-a_i h^2 - b_i h) - \frac{-d_i - d_{i+1}}{h}$$

$$c_i = -(\frac{M_{i+1} - M_i}{6h} h^2 + \frac{M_i}{2} h) - \frac{y_i - y_{i+1}}{h}$$

$$c_i = \frac{y_{i+1} - y_i}{h} - (\frac{M_{i+1} - M_i}{6} h + \frac{3M_i}{6} h)$$

$$c_i = \frac{y_{i+1} - y_i}{h} - (\frac{M_{i+1} - M_i + 3M_i}{6} h)$$

$$c_i = \frac{y_{i+1} - y_i}{h} - (\frac{M_{i+1} - 2M_i}{6}) h.$$

We now have our equations for determining the weights for our $n-1$ equations,

$$a_i = \frac{M_{i+1} - M_i}{6h}$$

$$b_i = \frac{M_i}{2}$$

$$c_i = \frac{y_{i+1} - y_i}{h} - (\frac{M_{i+1} + 2M_i}{6}) h$$

$$d_i = y_i$$

These systems can be handled more conveniently by putting them into matrix form as follows:

$$c_{i+1} = 3a_i h^2 + ab_i h + c_i$$

$$3(\frac{M_{i+1} - M_i}{6h})h^2 + 2(\frac{M_i}{2})h + \frac{y_{i+1} - y_i}{h} - (\frac{M_{i+1} + 2M_i}{6})h = \frac{y_{i+2} - y_{i+1}}{h} - (\frac{M_{i+2} - 2M_{i+1}}{6})h$$

$$3(\frac{M_{i+1} - M_i}{6h})h^2 + 2(\frac{M_i}{2})h - (\frac{M_{i+1} + 2M_i}{6})h + (\frac{M_{i+2} + 2M_{i+1}}{6})h = -(\frac{y_{i+1} - y_i}{h}) + \frac{y_{i+2} - y_{i+1}}{h}$$

$$h(\frac{3M_{i+1} - 3M_i}{6} + \frac{6M_i}{6} - (\frac{M_{i+1} + 2M_i}{6}) + (\frac{M_{i+2} + 2M_{i+1}}{6})) = \frac{y_i - 2y_{i+1} + y_{i+2}}{h}$$

$$\frac{h}{6}(M_i + 4M_{i+1} + M_{i+2}) = \frac{y_i - 2y_{i+1} + y_{i+2}}{h}$$

$$M_i + 4M_{i+1} + M_{i+2} = 6(\frac{y_i - 2y_{i+1} + y_{i+2}}{h^2})$$

for $i = 1, 2, 3, \ldots, n-1$

which leads to the matrix equation,

$$\begin{bmatrix} 1 & 4 & 1 & 0 & \cdots & 0 & 0 & 0 & 0 \\ 0 & 1 & 4 & 1 & \cdots & 0 & 0 & 0 & 0 \\ 0 & 0 & 1 & 4 & \cdots & 0 & 0 & 0 & 0 \\ \vdots & \vdots & \vdots & \vdots & \ddots & \vdots & \vdots & \vdots & \vdots \\ 0 & 0 & 0 & 0 & \cdots & 4 & 1 & 0 & 0 \\ 0 & 0 & 0 & 0 & \cdots & 1 & 4 & 1 & 0 \\ 0 & 0 & 0 & 0 & \cdots & 0 & 1 & 4 & 1 \end{bmatrix} \begin{bmatrix} M_1 \\ M_2 \\ M_3 \\ M_4 \\ \vdots \\ M_{n-3} \\ M_{n-2} \\ M_{n-1} \\ M_n \end{bmatrix} = \frac{6}{h^2} \begin{bmatrix} y_1 - 2y_2 + y_3 \\ y_2 - 2y_3 + y_4 \\ y_3 - 2y_4 + y_5 \\ \vdots \\ y_{n-4} - 2y_{n-3} + y_{n-2} \\ y_{n-3} - 2y_{n-2} + y_{n-1} \\ y_{n-2} - 2y_{n-1} + y_n \end{bmatrix}$$

Note that this system has $n-2$ rows and n columns, and is therefore under-determined. In order to generate a unique cubic spline, two other conditions must be imposed upon the system.

Three Types of Splines

Natural Splines

This first spline type includes the stipulation that the second derivative be equal to zero at the endpoints.

$$M_1 = M_n = 0$$

This results in the spline extending as a line outside the endpoints. The matrix for determining the $M_1 - M_n$ values can be adapted accordingly.

$$\begin{bmatrix} 1 & 0 & 0 & 0 & \cdots & 0 & 0 & 0 & 0 \\ 0 & 1 & 4 & 1 & \cdots & 0 & 0 & 0 & 0 \\ 0 & 0 & 1 & 4 & \cdots & 0 & 0 & 0 & 0 \\ \vdots & \vdots & \vdots & \vdots & \ddots & \vdots & \vdots & \vdots & \vdots \\ 0 & 0 & 0 & 0 & \cdots & 4 & 1 & 0 & 0 \\ 0 & 0 & 0 & 0 & \cdots & 1 & 4 & 1 & 0 \\ 0 & 0 & 0 & 0 & \cdots & 0 & 1 & 0 & 1 \end{bmatrix} \begin{bmatrix} 0 \\ M_2 \\ M_3 \\ M_4 \\ \vdots \\ M_{n-3} \\ M_{n-2} \\ M_{n-1} \\ 0 \end{bmatrix} = \frac{6}{h^2} \begin{bmatrix} y_1 - 2y_2 + y_3 \\ y_2 - 2y_3 + y_4 \\ y_3 - 2y_4 + y_5 \\ \vdots \\ y_{n-4} - 2y_{n-3} + y_{n-2} \\ y_{n-3} - 2y_{n-2} + y_{n-1} \\ y_{n-2} - 2y_{n-1} + y_n \end{bmatrix}$$

For reasons of convenience, the first and last columns of this matrix can be eliminated, as they correspond to the M_1 and M_n values, which are both 0.

$$\begin{bmatrix} 4 & 1 & 0 & \cdots & 0 & 0 & 0 \\ 1 & 4 & 1 & \cdots & 0 & 0 & 0 \\ 0 & 1 & 4 & \cdots & 0 & 0 & 0 \\ \vdots & \vdots & \vdots & \ddots & \vdots & \vdots & \vdots \\ 0 & 0 & 0 & \cdots & 4 & 1 & 0 \\ 0 & 0 & 0 & \cdots & 1 & 4 & 1 \\ 0 & 0 & 0 & \cdots & 0 & 1 & 4 \end{bmatrix} \begin{bmatrix} M_2 \\ M_3 \\ M_4 \\ \vdots \\ M_{n-3} \\ M_{n-2} \\ M_{n-1} \end{bmatrix} = \frac{6}{h^2} \begin{bmatrix} y_1 - 2y_2 + y_3 \\ y_2 - 2y_3 + y_4 \\ y_3 - 2y_4 + y_5 \\ \vdots \\ y_{n-4} - 2y_{n-3} + y_{n-2} \\ y_{n-3} - 2y_{n-2} + y_{n-1} \\ y_{n-2} - 2y_{n-1} + y_n \end{bmatrix}$$

This results in an $n-2$ by $n-2$ matrix, which will determine the remaining solutions for M_2 through M_{n-1}. The spline is now unique.

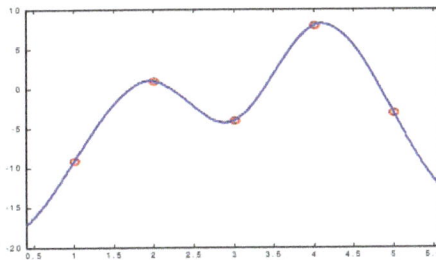

Figure: Natural interpolating curve

Parabolic Runout Spline

The parabolic spline imposes the condition that the second derivative at the endpoints, M_1 and M_n, be equal to M_2 and M_{n-1} respectively.

$$M_1 = M_2$$
$$M_n = M_{n-1}$$

The result of this condition is the curve becomes a parabolic curve at the endpoint. This type of cubic spline is useful for periodic and exponential data.

The matrix equation for this type of spline is,

$$
\begin{bmatrix}
5 & 1 & 0 & \cdots & 0 & 0 & 0 \\
1 & 4 & 1 & \cdots & 0 & 0 & 0 \\
0 & 1 & 4 & \cdots & 0 & 0 & 0 \\
\vdots & \vdots & \vdots & \ddots & \vdots & \vdots & \vdots \\
0 & 0 & 0 & \cdots & 4 & 1 & 0 \\
0 & 0 & 0 & \cdots & 1 & 4 & 1 \\
0 & 0 & 0 & \cdots & 0 & 1 & 5
\end{bmatrix}
\begin{bmatrix}
M_2 \\ M_3 \\ M_4 \\ \vdots \\ M_{n-3} \\ M_{n-2} \\ M_{n-1}
\end{bmatrix}
= \frac{6}{h^2}
\begin{bmatrix}
y_1 - 2y_2 + y_3 \\
y_2 - 2y_3 + y_4 \\
y_3 - 2y_4 + y_5 \\
\vdots \\
y_{n-4} - 2y_{n-3} + y_{n-2} \\
y_{n-3} - 2y_{n-2} + y_{n-1} \\
y_{n-2} - 2y_{n-1} + y_n
\end{bmatrix}
$$

We can now determine the values for M_2 through M_{n-1}, with the values for M_1 and M_n already determined.

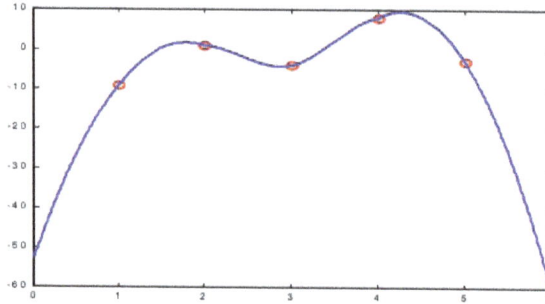

Figure: Parabolic Runout curve

Note that the endpoint behavior is a bit more extreme than with the natural spline option.

Cubic Runout Spline

This last type of spline has the most extreme endpoint behavior. It assigns M_1 to be $2M_2 - M_3$ and M_n to be $2M_{n-1} - M_{n-2}$. this cause the curve to degrade to a single cubic curve over the last two intervals, rather than two separate functions.

The matrix equation for this type is,

$$
\begin{bmatrix}
6 & 0 & 0 & \cdots & 0 & 0 & 0 \\
1 & 4 & 1 & \cdots & 0 & 0 & 0 \\
0 & 1 & 4 & \cdots & 0 & 0 & 0 \\
\vdots & \vdots & \vdots & \ddots & \vdots & \vdots & \vdots \\
0 & 0 & 0 & \cdots & 4 & 1 & 0 \\
0 & 0 & 0 & \cdots & 1 & 4 & 1 \\
0 & 0 & 0 & \cdots & 0 & 1 & 6
\end{bmatrix}
\begin{bmatrix}
M_2 \\ M_3 \\ M_4 \\ \vdots \\ M_{n-3} \\ M_{n-2} \\ M_{n-1}
\end{bmatrix}
= \frac{6}{h^2}
\begin{bmatrix}
y_1 - 2y_2 + y_3 \\
y_2 - 2y_3 + y_4 \\
y_3 - 2y_4 + y_5 \\
\vdots \\
y_{n-4} - 2y_{n-3} + y_{n-2} \\
y_{n-3} - 2y_{n-2} + y_{n-1} \\
y_{n-2} - 2y_{n-1} + y_n
\end{bmatrix}
$$

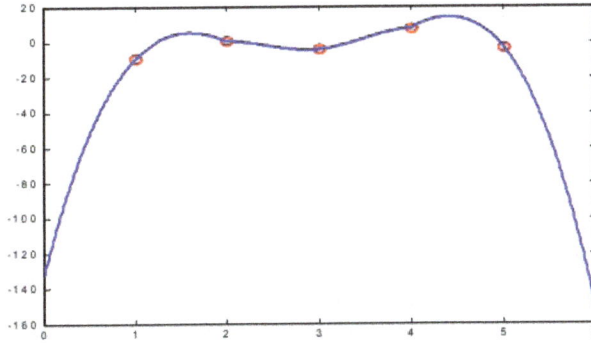

Figure: Cubic Runout curve

References

- Schatzman, Michelle (2002), "Chapter 4", Numerical Analysis: A Mathematical Introduction, Oxford: Clarendon Press, ISBN 0-19-850279-6

- Interpolation, science: britannica.com, Retrieved 14 May 2018

- Waring, Edward (9 January 1779). "Problems concerning interpolations" (PDF). Philosophical Transactions of the Royal Society. 69: 59–67. doi:10.1098/rstl.1779.0008

- Polynomial-interpolation: techtarget.com, Retrieved 27 April 2018

- Quarteroni, Alfio; Saleri, Fausto (2003). Scientific Computing with MATLAB. Texts in computational science and engineering. 2. Springer. p. 66. ISBN 978-3-540-44363-6

- Lagrange-interpolation: brilliant.org, Retrieved 25 June 2018

- Meijering, Erik (2002). "A chronology of interpolation: from ancient astronomy to modern signal and image processing" (PDF). Proceedings of the IEEE. 90 (3): 319–342. doi:10.1109/5.993400

Numerical Integration

Numerical integration comprises a family of algorithms that are meant for the calculation of the numerical value of a definite integral. It also encompasses the solution of differential equations using numerical methods. The topics elaborated in this chapter will help in developing a better perspective about truncation error, mid point rule, trapezoidal rule, Simpson's rule, Newton–Cotes formulas and Gaussian quadrature, for a comprehensive understanding of numerical integration.

Numerical Integration is the process of computing the value of a definite integral, $\int_a^b f(x)dx$, when the values of the integrand function, $y = f(x)$ are given at some tabular points.

Elements of Numerical Integration

If f(x) is a smooth wellbehaved function, integrated over a small number of dimensions and the limits of integration are bounded, there are many methods of approximating the integral with arbitrary precision. We consider an indefinite integral:

$$\int_a^b f(x)\,dx.$$

Numerical integration methods can generally be described as combining evaluations of the integrand to get an approximation to the integral. The integrand is evaluated at a finite set of points called integration points and a weighted sum of these values is used to approximate the integral. For instance if we use rectangles as our shape:

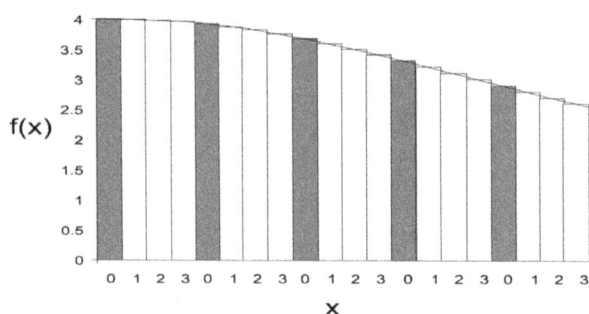

In this example the definite integral is thus approximated using areas of rectangles. The integration points and weights depend on the specific method used and the accuracy required from the approximation. An important part of the analysis of any numerical integration method is to study the behavior of the approximation error as a function of the number of integrand evaluations. A method which yields a small error for a small number of evaluations is usually considered superior. Reducing the number of evaluations of the integrand reduces the number of arithmetic operations involved, and therefore reduces the total roundoff error. Also, each evaluation takes time, and the integrand may be arbitrarily complicated.

Note that if one were to take an infinite number of divisions this would approach the analytical function (derived in calculus) representing the area under the curve. We do not do this in practice as an infinite number of divisions would require a prohibitively expensive amount of computing power is rarely ever needed to be exact.

Methods for One-dimensional Integrals

Numerical integration methods can generally be described as combining evaluations of the integrand to get an approximation to the integral. The integrand is evaluated at a finite set of points called integration points and a weighted sum of these values is used to approximate the integral. The integration points and weights depend on the specific method used and the accuracy required from the approximation.

An important part of the analysis of any numerical integration method is to study the behavior of the approximation error as a function of the number of integrand evaluations. A method that yields a small error for a small number of evaluations is usually considered superior. Reducing the number of evaluations of the integrand reduces the number of arithmetic operations involved, and therefore reduces the total round-off error. Also, each evaluation takes time, and the integrand may be arbitrarily complicated.

A 'brute force' kind of numerical integration can be done, if the integrand is reasonably well-behaved (i.e. piecewise continuous and of bounded variation), by evaluating the integrand with very small increments.

Quadrature Rules based on Interpolating Functions

A large class of quadrature rules can be derived by constructing interpolating functions that are easy to integrate. Typically these interpolating functions are polynomials. In practice, since polynomials of very high degree tend to oscillate wildly, only polynomials of low degree are used, typically linear and quadratic.

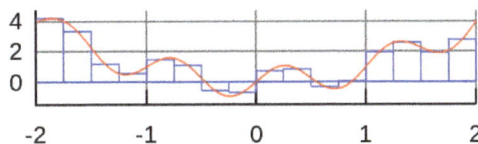

The rectangle rule

The simplest method of this type is to let the interpolating function be a constant function (a polynomial of degree zero) that passes through the point $\left(\dfrac{a+b}{2}, f\left(\dfrac{a+b}{2}\right)\right)$. This is called the *midpoint rule* or *rectangle rule*.

$$\int_a^b f(x)dx \approx (b-a)f\left(\frac{a+b}{2}\right).$$

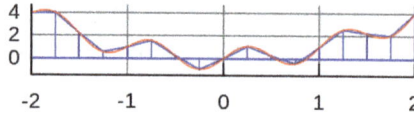

The trapezoidal rule

The interpolating function may be a straight line (an affine function, i.e. a polynomial of degree 1) passing through the points $(a, f(a))$ and $(b, f(b))$. This is called the *trapezoidal rule*.

$$\int_a^b f(x)dx \approx (b-a)\left(\frac{f(a)+f(b)}{2}\right).$$

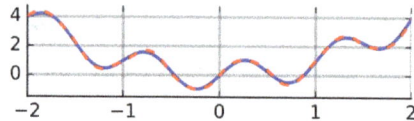

Simpson's rule

For either one of these rules, we can make a more accurate approximation by breaking up the interval $[a,b]$ into some number n of subintervals, computing an approximation for each subinterval, then adding up all the results. This is called a *composite rule*, *extended rule*, or *iterated rule*. For example, the composite trapezoidal rule can be stated as,

$$\int_a^b f(x)dx \approx \frac{b-a}{n}\left(\frac{f(a)}{2} + \sum_{k=1}^{n-1}\left(f\left(a+k\frac{b-a}{n}\right)\right) + \frac{f(b)}{2}\right),$$

where the subintervals have the form $[a+kh, a+(k+1)h] \subset [a,b]$, with $h = \dfrac{b-a}{n}$ and $k = 0,\ldots,n-1$. Here we used subintervals of the same length h but one could also use intervals of varying length $(h_k)_k$.

Interpolation with polynomials evaluated at equally spaced points in $[a,b]$ yields the Newton–Cotes formulas, of which the rectangle rule and the trapezoidal rule are examples. Simpson's rule, which is based on a polynomial of order 2, is also a Newton–Cotes formula.

Quadrature rules with equally spaced points have the very convenient property of *nesting*. The corresponding rule with each interval subdivided includes all the current points, so those integrand values can be re-used.

If we allow the intervals between interpolation points to vary, we find another group of quadrature formulas, such as the Gaussian quadrature formulas. A Gaussian quadrature rule is typically more accurate than a Newton–Cotes rule, which requires the same number of function evaluations, if the integrand is smooth (i.e., if it is sufficiently differentiable). Other quadrature methods with varying intervals include Clenshaw–Curtis quadrature (also called Fejér quadrature) methods, which do nest.

Gaussian quadrature rules do not nest, but the related Gauss–Kronrod quadrature formulas do.

Adaptive Algorithms

If $f(x)$ does not have many derivatives at all points, or if the derivatives become large, then Gaussian quadrature is often insufficient. In this case, an algorithm similar to the following will perform better:

```
def calculate_definite_integral_of_f(f, initial_step_size):
    '''
    This algorithm calculates the definite integral of a function
    from 0 to 1, adaptively, by choosing smaller steps near
    problematic points.
    '''
    x = 0.0
    h = initial_step_size
    accumulator = 0.0
    while x < 1.0:
        if x + h > 1.0:
            h = 1.0 - x # At end of unit interval, adjust last step to end
at 1.
        if error_too_big_in_quadrature_of_f_over_range(f, [x,x+h]):
            h = make_h_smaller(h)
        else:
            accumulator += quadrature_of_f_over_range(f, [x,x+h])
            x += h
            if error_too_small_in_quadrature_of_over_range(f, [x,x+h]):
                h = make_h_larger(h) # Avoid wasting time on tiny steps.
    return accumulator
```

Some details of the algorithm require careful thought. For many cases, estimating the error from quadrature over an interval for a function $f(x)$ isn't obvious. One popular solution is to use two different rules of quadrature, and use their difference as an estimate of the error from quadrature. The other problem is deciding what "too large" or "very small" signify. A *local* criterion for "too large" is that the quadrature error should not be larger than $t \cdot h$ where t, a real number, is the tolerance we wish to set for global error. Then again, if h is already tiny, it may not be worthwhile to make it even smaller even if the quadrature error is apparently large. A *global* criterion is that the sum of errors on all the intervals should be less than t. This type of error analysis is usually called "a posteriori" since we compute the error after having computed the approximation.

Extrapolation Methods

The accuracy of a quadrature rule of the Newton-Cotes type is generally a function of the number of evaluation points. The result is usually more accurate as the number of evaluation points increases, or, equivalently, as the width of the step size between the points decreases. It is natural to ask what the result would be if the step size were allowed to approach zero. This can be answered by extrapolating the result from two or more nonzero step sizes, using series acceleration methods such as Richardson extrapolation. The extrapolation function may be a polynomial or rational function. Extrapolation methods are described in more detail by Stoer and Bulirsch and are implemented in many of the routines in the QUADPACK library.

Conservative (a Priori) Error Estimation

Let f have a bounded first derivative over $[a,b]$, i.e. $f \in C^1([a,b])$. The mean value theorem for f, where $x \in [a,b)$, gives

$$(x-a)f'(\xi_x) = f(x) - f(a),$$

for some $\xi_x \in (a,x]$ depending on x. If we integrate in x from a to b on both sides and take the absolute values, we obtain

$$\left| \int_a^b f(x)dx - (b-a)f(a) \right| = \left| \int_a^b (x-a)f'(v_x)dx \right|.$$

We can further approximate the integral on the right-hand side by bringing the absolute value into the integrand, and replacing the term in f' by an upper bound

$$\left| \int_a^b f(x)dx - (b-a)f(a) \right| \le \frac{(b-a)^2}{2} \sup_{a \le x \le b} |f'(x)|,$$

where the supremum was used to approximate. Hence, if we approximate the integral $\int_a^b f(x)dx$ by the quadrature rule $(b-a)f(a)$ our error is no greater than the right hand side of 1. We can convert this into an error analysis for the Riemann sum (*), giving an upper bound of,

$$\frac{n^{-1}}{2}\sup_{0\le x\le 1}|f'(x)|$$

for the error term of that particular approximation. (Note that this is precisely the error we calculated for the example $f(x)=x$.) Using more derivatives, and by tweaking the quadrature, we can do a similar error analysis using a Taylor series (using a partial sum with remainder term) for f. This error analysis gives a strict upper bound on the error, if the derivatives of f are available.

This integration method can be combined with interval arithmetic to produce computer proofs and *verified* calculations.

Integrals Over Infinite Intervals

Several methods exist for approximate integration over unbounded intervals. The standard technique involves specially derived quadrature rules, such as Gauss-Hermite quadrature for integrals on the whole real line and Gauss-Laguerre quadrature for integrals on the positive reals. Monte Carlo methods can also be used, or a change of variables to a finite interval; e.g., for the whole line one could use,

$$\int_{-\infty}^{\infty} f(x)dx = \int_{-1}^{+1} f\left(\frac{t}{1-t^2}\right)\frac{1+t^2}{(1-t^2)^2}dt,$$

and for semi-infinite intervals one could use,

$$\int_a^{\infty} f(x)dx = \int_0^1 f\left(a+\frac{t}{1-t}\right)\frac{dt}{(1-t)^2},$$

$$\int_{-\infty}^a f(x)dx = \int_0^1 f\left(a-\frac{1-t}{t}\right)\frac{dt}{t^2},$$

as possible transformations.

Multidimensional Integrals

The quadrature rules discussed so far are all designed to compute one-dimensional integrals. To compute integrals in multiple dimensions, one approach is to phrase the multiple integral as repeated one-dimensional integrals by applying to Fubini's theorem. This approach requires the function evaluations to grow exponentially as the number of dimensions increases. Three methods are known to overcome this so-called *curse of dimensionality*.

Monte Carlo

Monte Carlo methods and quasi-Monte Carlo methods are easy to apply to multi-dimensional integrals, and may yield greater accuracy for the same number of function evaluations than repeated integrations using one-dimensional methods.

A large class of useful Monte Carlo methods are the so-called Markov chain Monte Carlo algorithms, which include the Metropolis-Hastings algorithm and Gibbs sampling.

Sparse Grids

Sparse grids were originally developed by Smolyak for the quadrature of high-dimensional functions. The method is always based on a one-dimensional quadrature rule, but performs a more sophisticated combination of univariate results.

Bayesian Quadrature

Bayesian Quadrature is a statistical approach to the numerical problem of computing integrals and falls under the field of probabilistic numerics. It can provide a full handling of the uncertainty over the solution of the integral expressed as a Gaussian Process posterior variance. It is also known to provide very fast convergence rates which can be up to exponential in the number of quadrature points n.

Connection with Differential Equations

The problem of evaluating the integral,

$$F(x) = \int_a^x f(u)du$$

can be reduced to an initial value problem for an ordinary differential equation by applying the first part of the fundamental theorem of calculus. By differentiating both sides of the above with respect to the argument x, it is seen that the function F satisfies,

$$\frac{dF(x)}{dx} = f(x), \quad F(a) = 0.$$

Methods developed for ordinary differential equations, such as Runge–Kutta methods, can be applied to the restated problem and thus be used to evaluate the integral. For instance, the standard fourth-order Runge–Kutta method applied to the differential equation yields Simpson's rule from above.

The differential equation $F'(x) = f(x)$ has a special form: the right-hand side contains only the dependent variable (here x) and not the independent variable (here F). This simplifies the theory and algorithms considerably. The problem of evaluating integrals is thus best studied in its own right.

Truncation Error

Truncation error is the difference between a truncated value and the actual value. A truncated quantity is represented by a numeral with a fixed number of allowed digits, with any excess digits "chopped off" (hence the expression "truncated").

As an example of truncation error, consider the speed of light in a vacuum. The official value is 299,792,458 meters per second. In scientific (power-of-10) notation, that quantity is expressed as 2.99792458×10^8. Truncating it to two decimal places yields 2.99×10^8. The truncation error is the difference between the actual value and the truncated value, or 0.00792458×10^8. Expressed properly in scientific notation, it is 7.92458×10^5.

Often an approximation is used in place of an exact mathematical procedure. For instance consider the Taylor series expansion of say $sinx$ i.e.

$$sinx = x - \frac{x^3}{3!} + \frac{x^5}{5!} - \frac{x^7}{7!} + ...$$

Practically we cannot use all of the infinite number of terms in the series for computing the sine of angle x. We usually terminate the process after a certain number of terms. The error that results due to such a termination or truncation is called as 'truncation error'.

Usually in evaluating logarithms, exponentials, trigonometric functions, hyperbolic functions etc. an infinite series of the form $S = \sum_{i=0}^{\infty} a_i x^i$ is replaced by a finite series $\sum_{i=0}^{n} a_i x_i$. Thus a truncation error of $\sum_{i=n+1}^{\infty} a_i x^i$ is is introduced in the computation.

For example let us consider evaluation of exponential function using first three terms at $x = 0.2$.

$$e^x = 1 + x + \frac{x^2}{2!} + \frac{x^3}{3!} + \frac{x^4}{4!} + \frac{x^5}{5!} + \frac{x^6}{6!} + ...$$

$$e^x \simeq 1 + x + \frac{x^2}{2!}$$

$$e^{0.2} \simeq 1 + 0.2 + \frac{0.04}{2} = 1.22$$

$$\text{Truncation Error} = \sum_{i=3}^{\infty} \frac{x^i}{i!} = \frac{x^3}{3!} + \frac{x^4}{4!} + \frac{x^5}{5!} + \frac{x^6}{6!} + \dots$$

$$= \frac{0.008}{6} + \frac{0.0016}{24} + \dots$$

$$= 0.001\overline{33} + 0.0000\overline{66} + \dots.$$

$$= 0.1\overline{33} \times 10^{-2} + 0.00\overline{66} \times 10^{-2} \dots$$

$$\therefore Truncation Error \leq 10^{-2}$$

Truncation Errors in Finite Difference Formulas

The accuracy of a finite difference formula is a fundamental issue when discretizing differential equations. We shall first go through a particular example in detail and thereafter list the truncation error in the most common finite difference approximation formulas.

Example: The backward difference for u′(t)

Consider a backward finite difference approximation of the first-order derivative u′:

$$[D_t^- u]^n = \frac{u^n - u^{n-1}}{\Delta t} \approx u'(t_n).$$

Here, u^n means the value of some function $u(t)$ at a point t_n, and $[D_t^- u]^n$ is the discrete derivative of $u(t)$ at $t=t_n$. The discrete derivative computed by a finite difference is not exactly equal to the derivative $u'(t_n)$. The error in the approximation is,

$$R^n = [D_t^- u]^n - u'(t_n).$$

The common way of calculating R^n s to:

1. Expand $u(t)$ in a Taylor series around the point where the derivative is evaluated, here t_n,

2. Insert this Taylor series in above equation, and

3. Collect terms that cancel and simplify the expression.

The result is an expression for R^n in terms of a power series in Δt. The error R^n is commonly referred to as the truncation error of the finite difference formula. The Taylor series formula often found in calculus books takes the form,

$$f(x+h) = \sum_{i=0}^{\infty} \frac{1}{i!} \frac{d^i f}{dx^i}(x) h^i.$$

In our application, we expand the Taylor series around the point where the finite differ-ence formula approximates the derivative. The Taylor series of u^n at t_n is simply $u(t_n)$, while the Taylor sereis of u^{n-1} at t_n must employ the general formula,

$$u(t_{n-1}) = u(t - \Delta t) = \sum_{i=0}^{\infty} \frac{1}{i!} \frac{d^i u}{dt^i}(t_n)(-\Delta t)^i$$

$$= u(t_n) - u'(t_n)\Delta t + \frac{1}{2}u''(t_n)\Delta t^2 + \mathcal{O}(\Delta t^3),$$

where $\mathcal{O}(\Delta t^3)$ means a power-series in Δt where the lowest power is Δt^3. We assume that Δt is small such that $\Delta t^p \gg \Delta t^q$ if p is smaller than q. The details of higher-order terms in Δt are therefore not of much interest. Inserting the Taylor series above in the left-hand side of 1 $R^n = [D_t^- u]^n - u'(t_n)$. gives rise to some algebra:

$$[D_t^- u]^n - u'(t_n) = \frac{u(t_n) - u(t_{n-1})}{\Delta t} - u'(t_n)$$

$$= \frac{u(t_n) - u(t_n) - u'(t_n)\Delta t + \frac{1}{2}u''(t_n)\Delta t^2 + \mathcal{O}(\Delta t^3))}{\Delta t} - u'(t_n)$$

$$= -\frac{1}{2}u''(t_n)\Delta t^2 + \mathcal{O}(\Delta t^2)),$$

which is, according to ($R^n = [D_t^- u]^n - u'(t_n)$.), the truncation error:

$$Rn = -\frac{1}{2}u''(t_n)\Delta t + O(\Delta t^2)).$$

The dominating term for small Δt is $-12u''(t_n)\Delta t$, which is proportional to Δt, and we say that the truncation error is of first order in Δt.

Example: The forward difference for u'(t)

We can analyze the approximation error in the forward difference

$$u'(t_n) \approx [D_t^+ u]^n = \frac{u^{n+1} - u^n}{\Delta t},$$

by writing

$$R^n = [D_t u]^n - u'(t_n),$$

and expanding u^{n+1} in a Taylor series around t_n,

$$u(t_{n+1}) = u(t_n) + u'(t_n)\Delta t + \frac{1}{2}u''(t_n)\Delta t^2 + \mathcal{O}(\Delta t^3).$$

The result becomes,

$$R = \frac{1}{2}u''(t_n)\,\Delta t + \mathcal{O}(\Delta t^2),$$

showing that also the forward difference is of first order.

Mid Point Rule

A point which is much more likely to be close to the average would be the midpoint of each subinterval. Using the midpoint in the sum is called the midpoint rule.

This rule uses the midpoint m_k of each of the intervals as the point at which to evaluate the function for the Riemann sum.

To estimate $\int_a^b f(t)dt$, the midpoint formula M_n with n slivers of equal width is:

$$M_n = \frac{b-a}{n}(f(m_1)+f(m_2)+...f(m_n)),$$

$$m_k = \frac{t_k+t_{k-1}}{2} = a+\frac{2k-1}{2n}(b-a).$$

Geometrically the area of the k-th rectangle, which is $\dfrac{b-a}{n}f(m_k)$, where m_k is the midpoint of the -sliver, can be viewed also as the area of the tangent trapezoid: this is the trapezoid of width $\dfrac{b-a}{n}$ and central height $f(m_k)$, which is tangent at the point $(m_k, f(m_k)$ to the graph of $y = f(t)$:

To see this we first note that the equation of the tangent line at $(m_k, f(m_k)$ is $y = f(m_k)+f'(m_k)(t-m_k)$. Then the left end of the tangent trapezoid (at $t = t_{k-1}$) has height:

$$L_k = f(m_k)+f'(m_k)(t_{k-1}-m_k),$$

and the right end (at $t = t_k$)) has height:

$$R_k = f(m_k)+f'(m_k)(t_k-m_k).$$

So the tangent trapezoid has area A_k given as follows:

$$A_k = \frac{1}{2}(b-a)(L_k+R_k)$$

$$= \frac{1}{2}(b-a)(2f(m_k) + f'(m_k)(t_{k-1} - m_k + t_k - m_k))$$

$$= (b-a)f(m_k),$$

since we have $t_{k-1} + t_k = 2m_k$, so the terms in A_k involving $f'(m_k)$ exactly cancel. So the area of the tangent trapezoid at m_k is the same as the area of the rectangle of height $f(m_k)$ and width $\frac{b-a}{n}$ which is exactly the k-th term of M_n. So the midpoint rule M_n estimates the area as the sum of the areas of the n-tangent trapezoids. How does M_n relate to the actual integral?

- If the graph of $y = f(t)$ is concave down, so in particular if $f'' \leq 0$, then each the tangent trapezoid lies above the graph of the function $y = f(t)$, so

$$M_n \geq \int_a^b f(t)dt.$$

- If the graph of $y = f(t)$, is concave up, so in particular if $f'' \geq 0$, then each tangent trapezoid lies below the graph of the function $y = f(t)$, so $M_n \leq \int_a^b f(t)dt$.

- In general, it can be shown that if $-K_2 \leq f''(t) \leq K_2$ on the interval $[a,b]$, then the error $E_n(M)$ in M_n (i.e. $E_n(M) = | M_n - \int_a^b f(t)dt |$ is bounded by the formula:

$$E_n \leq \frac{K_2}{24n^2}(b-a)^3.$$

Examples:

- The integral $\int_1^3 \frac{1}{t}dt$ is estimated by the midpoint rule with 6 intervals as:

$$M_6 = \frac{3-1}{6}(f(\frac{7}{6}) + f(\frac{3}{2}) + f(\frac{11}{6}) + f(\frac{13}{6}) + f(\frac{5}{2}) + f(\frac{17}{6}))$$

$$= 2(\frac{1}{7} + \frac{1}{9} + \frac{1}{11} + \frac{1}{13} + \frac{1}{15} + \frac{1}{17})$$

$$= \frac{838192}{765765} = 1.094581236.$$

Here $f(t) = \frac{1}{t}$.

The error estimate goes as follows:

$$f'(t) = -\frac{1}{t^2}, f''(t) = \frac{2}{t^3}.$$

On the interval we have $f''(t) > 0$, so the graph of $y = f(t)$ is concave up and the midpoint estimate is too low.

The maximum of $|f''(t)|$ is attained at $t = 1$ and is 2, So we may take $K_2 = 2$ and then we have:

$$E_6(M) \leq \frac{K}{24n^2}(b-a)^3 = \frac{2}{24(6^2)}(3-1)^3 = \frac{1}{432}2^3 = \frac{1}{54}.$$

So the maximum possible error is $\dfrac{1}{54}$. The actual error is $\ln(3) - \dfrac{838192}{765765} = 0.004031053$, which is well within the maximum possible error.

Note that combining the estimate M_6 and the estimate T_6 done earlier, we have the estimate:

$$1.094581236 \leq \ln(3) \leq 1.106746032.$$

Thus by doing both M_6 and T_6, we get quite a good estimate for $\ln(3)$:

$$\frac{1}{2}\left(\frac{838192}{765765} + \frac{2789}{2520}\right) = \frac{2697119}{2450448} = 1.100663634,$$

where the error is at most:

$$\frac{1}{2}\left(\frac{838192}{765765} - \frac{2789}{2520}\right) = \frac{24841}{4084080} = .006082397994.$$

The actual error is now just 0.002051345, less than two parts in a thousand.

- The integral $\int_1^4 e^t dt$ is estimated by the midpoint rule with 6 intervals as:

$$M_6 = \frac{4-1}{6}(e^{\frac{5}{4}} + e^{\frac{7}{4}} + e^{\frac{9}{4}} + e^{\frac{11}{4}} + e^{\frac{13}{4}} + e^{\frac{15}{4}})$$

$$= \frac{1}{2}e^{\frac{5}{4}}\frac{e^3 - 1}{e^{\frac{1}{2}} - 1} = 51.34336765.$$

The second derivative of e^t is e^t which has its maximum on the interval at $t = 4$. Since the second derivative is positive, the graph of $y = e^t$ is concave up and the midpoint rule gives an underestimate. So for K_2 we may take e^4 and the maximum error is:

$$E_6(M) \leq \frac{e^4}{24(6^2)}(4-1)^3 = \frac{e^4}{32} = 1.7063192188.$$

The exact value of the integral is $e^4 - e = 51.87986820$, so the true error is about 0.53650055, which is well within the allowed error.

Note that combining the estimate   and the estimate T_6 done earlier, we have the estimate:

$$51.34336765 \le \int_1^4 e^t dt \le 52.95622196.$$

Thus by doing both M_6 and T_6, we get quite a good estimate for $\int_1^4 e^t dt$:

$$\frac{1}{2}(52.95622196 + 51.34336765) = 52.14979480,$$

where the error is at most:

$$\frac{1}{2}(52.95622196 - 51.34336765) = 0.8064271550.$$

The actual error is now just 0.26992660, less than six parts in a thousand.

Trapezoidal Rule

In mathematics, and more specifically in numerical analysis, the trapezoidal rule, also known as the trapezoid rule or trapezium rule, is a technique for approximating the definite integral.

The trapezoidal rule works by approximating the region under the graph of the function f(x) as a trapezoid and calculating its area.

The trapezoidal rule is to find the exact value of a definite integral using a numerical method. This rule is mainly based on the Newton-Cotes formula which states that one can find the exact value of the integral as an nth order polynomial.

When n = 1 according to Trapezoidal rule, the area under the linear polynomial is stated as,

$$\int_a^b f(x)dx = (b-a)\left[\frac{f(a)+f(b)}{2}\right]$$

Derivation of the Trapezoidal Rule

Method 1: Derived from Calculus

$$\int_a^b f(x)dx \approx \int_a^b f_1(x)dx$$

$$= \int_a^b (a_o + a\,x)dx$$

$$= a_o(b-a) + a_1\left(\frac{b^2 - a^2}{2}\right)$$

But what is a_o and a_1? Now if one chooses, $(a, f(a))$ and $(b, f(b))$ as the two points to approximate $f(x)$ by a straight line from a to b,

$$f(a) = f_1(a) = a_o + a_1 a$$
$$f(b) = f_1(b) = a_o + a_1 b$$

Solving the above two equations for a_1 and a_o,

$$a_1 = \frac{f(b) - f(a)}{b - a}$$

$$a_o = \frac{f(a)\,b - f(b)\,a}{b - a}$$

Hence from Equation $= a_o(b-a) + a_1\left(\frac{b^2 - a^2}{2}\right)$,

$$\int_a^b f(x)dx \approx \frac{f(a)\,b - f(b)\,a}{b-a}(b-a) + \frac{f(b) - f(a)}{b-a}\,\frac{b^2 - a^2}{2}$$

$$= (b-a)\left[\frac{f(a) - f(b)}{2}\right]$$

Method 2: Also Derived from Calculus

$f_1(x)$ can also be approximated by using Newton's divided difference polynomial as

$$f_1(x) = f(a) + \frac{f(b) - f(a)}{b - a}(x - a)$$

Hence

$$\int_a^b f(x)dx \approx \int_a^b f_1(x)dx$$

$$= \int_a^b \left[f(a) + \frac{f(b) - f(a)}{b - a}(x - a) \right] dx$$

$$= \left[f(a)x + \frac{f(b)-f(a)}{b-a}\left(\frac{x^2}{2} - ax\right) \right]_a^b$$

$$= f(a)b - f(a)a + \left(\frac{f(b)-f(a)}{b-a}\right)\left(\frac{b^2}{2} - ab - \frac{a^2}{2} + a^2\right)$$

$$= f(a)b - f(a)a + \left(\frac{f(b)-f(a)}{b-a}\right)\left(\frac{b^2}{2} - ab + \frac{a^2}{2}\right)$$

$$= f(a)b - f(a)a + \left(\frac{f(b)-f(a)}{b-a}\right)\frac{1}{2}(b-a)^2$$

$$= f(a)b - f(a)a + \frac{1}{2}(f(b)-f(a))(b-a)$$

$$= f(a)b - f(a)a + \frac{1}{2}f(b)b - \frac{1}{2}f(b)a - \frac{1}{2}f(a)b + \frac{1}{2}f(a)a$$

$$= \frac{1}{2}f(a)b - \frac{1}{2}f(a)a + \frac{1}{2}f(b)b - \frac{1}{2}f(b)a$$

$$= (b-a)\left[\frac{f(a)+f(b)}{2}\right]$$

This gives the same result as Method two equation because they are just different forms of writing the same polynomial.

Method 3: Derived from Geometry

The trapezoidal rule can also be derived from geometry. Look at figure below. The area under the curve $f_1(x)$ is the area of a trapezoid. The integral:

$$\int_a^b f(x)dx \approx \text{Area of trapezoid}$$

$$= \frac{1}{2} \text{ (Sum of length of parallel sides) (Perpendicular distance between parallel sides)}$$

$$\frac{1}{2}(f(b)+f(a))(b-a)$$

$$= (b-a)\left[\frac{f(a)+f(b)}{2}\right]$$

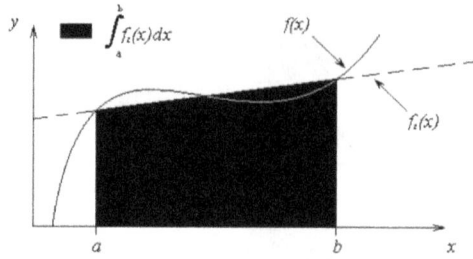

Figure: Geometric representation of trapezoidal rule

Method 4: Derived from Method of Coefficients

The trapezoidal rule can also be derived by the method of coefficients. The formula,

$$\int_a^b f(x)dx \approx \frac{b-a}{2}f(a) + \frac{b-a}{2}f(b)$$

$$= \sum_{i=1}^{2} c_i f(x_i)$$

Where

$$\frac{b \quad a}{}$$

$$c_2 = \frac{b-a}{2}$$

$$x_1 = a$$

$$x_1 = b$$

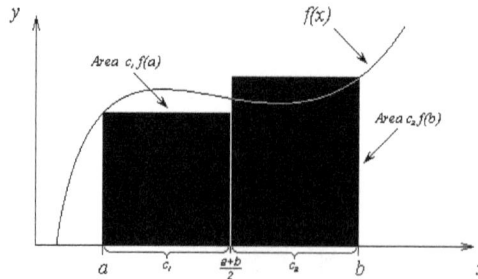

Figure: Area by method of coefficients

The interpretation is that $f(x)$ is evaluated at points a and b , and each function evaluation is given a weight of $\frac{b-a}{2}$. Geometrically, Method three equation is looked at as the area of a trapezoid, while Method four equation is viewed as the sum of the area of two rectangles, as shown in figure below. How can one derive the trapezoidal rule by the method of coefficients?

Assume,

$$\int_a^b f(x)dx = c_1 f(a) + c_2 f(b)$$

Let the right hand side be an exact expression for integrals of $\int_a^b 1dx$ and $\int_a^b xdx$, that is,

the formula will then also be exact for linear combinations of $f(x) = 1$ and $f(x) = x$, that is, for $f(x) = a_0(1) + a_1(x)$.

$$\int_a^b 1dx = b - a = c_1 + c_2$$

$$\int_a^b xdx = \frac{b^2 - a^2}{2} = c_1 a + c_2 b$$

Solving the above two equations gives,

$$c_1 = \frac{b-a}{2}$$

$$c_2 = \frac{b-a}{2}$$

Hence,

$$\int_a^b f(x)dx \approx \frac{b-a}{2} f(a) + \frac{b-a}{2} f(b)$$

Method 5: Another approach on the Method of Coefficients,

The trapezoidal rule can also be derived by the method of coefficients by another approach,

$$\int_a^b f(x)dx \approx \frac{b-a}{2} f(a) + \frac{b-a}{2} f(b)$$

Assume,

$$\int_a^b f(x)dx = c_1 f(a) + c_2 f(b)$$

Let the right hand side be exact for integrals of the form,

$$\int_a^b (a_0 + a_1 x)dx$$

So

$$\int_a^b (a_0 + a_1 x)dx = \left(a_0 x + a_1 \frac{x^2}{2} \right)_a^b$$

$$= a_0(b-a) + a_1 \left(\frac{b^2 - a^2}{2} \right)$$

But we want,

$$\int_a^b (a_0 + a_1 x)dx = c_1 f(a) + c_2 f(b)$$

to give the same result as Equation $\left(= a_0(b-a) + a_1 \left(\dfrac{b^2 - a^2}{2} \right) \right)$ for $f(x) = a_0 + a_1 x$.

$$\int_a^b (a_0 + a_1 x)dx = c_1(a_0 + a_1 a) + c_2(a_0 + a_1 b)$$

$$= a_0(c_1 + c_2) + a_1(c_1 a + c_2 b)$$

Hence from equations $= a_0(b-a) + a_1 \left(\dfrac{b^2 - a^2}{2} \right)$ and $(a_0(c_1 + c_2) + a_1(c_1 a + c_2 b))$,

$$= a_0(b-a) + a_1 \left(\frac{b^2 - a^2}{2} \right) = a_0(c_1 + c_2) + a_1(c_1 a + c_2 b)$$

Since a_0 and a_1 are arbitrary for a general straight line,

$$c_1 + c_2 = b - a$$

$$c_1 a + c_2 b = \frac{b^2 - a^2}{2}$$

Again, solving the above two equations gives,

$$c_1 = \frac{b - a}{2}$$

$$c_2 = \frac{b - a}{2}$$

Therefore,

$$\int_a^b f(x)dx \approx c_1 f(a) + c_2 f(b)$$

$$= \frac{b-a}{2}f(a) + \frac{b-a}{2}f(b)$$

Error Analysis

The error of the composite trapezoidal rule is the difference between the value of the integral and the numerical result:

$$\text{error} = \int_a^b f(x)dx - \frac{b-a}{N}\left[\frac{f(a)+f(b)}{2} + \sum_{k=1}^{N-1} f\left(a+k\frac{b-a}{N}\right)\right]$$

There exists a number ξ between a and b, such that,

$$\text{error} = -\frac{(b-a)^3}{12N^2}f''(\xi)$$

It follows that if the integrand is concave up (and thus has a positive second derivative), then the error is negative and the trapezoidal rule overestimates the true value. This can also be seen from the geometric picture: the trapezoids include all of the area under the curve and extend over it. Similarly, a concave-down function yields an underestimate because area is unaccounted for under the curve, but none is counted above. If the interval of the integral being approximated includes an inflection point, the error is harder to identify.

Proof

First suppose that $h = \frac{b-a}{N}$, $a_k = a+(k-1)h$, and

$$g_k(t) = \frac{1}{2}t[f(a_k)+f(a_k+t)] - \int_{a_k}^{a_k+t} f(x)dx$$

Then:

$$\frac{dg_k}{dt} = \frac{1}{2}[f(a_k)+f(a_k+t)] + \frac{1}{2}t\cdot f'(a_k+t) - f(a_k+t)$$

$$\frac{d^2g_k}{dt^2} = \frac{1}{2}t\cdot f''(a_k+t)$$

and suppose that $|f''(x)| \leq f''(\xi)$, so that:

$$\left|f''(a_k+t)\right| \le f''(\xi), \ -f''(\xi) \le f''(a_k+t) \le f''(\xi)$$

$$-\frac{f''(\xi)t}{2} \le g''_k(t) \le \frac{f''(\xi)t}{2}$$

Because of $g'_k(0)=0$, $g_k(0)=0$

$$\int_0^t g''_k(x)dx = g''_k(t), \ \int_0^t g'_k(x)dx = g_k(t)$$

so that:

$$-\frac{f''(\xi)t^2}{4} \le g'_k(t) \le \frac{f''(\xi)t^2}{4}$$

$$-\frac{f''(\xi)t^3}{12} \le g_k(t) \le \frac{f''(\xi)t^3}{12}$$

Then assign h to t.

$$-\frac{f''(\xi)h^3}{12} \le g_k(h) \le \frac{f''(\xi)h^3}{12}$$

and:

$$\sum_{k=1}^{N} g_k(h) = \frac{b-a}{N}\left[\frac{f(a)+f(b)}{2} + \sum_{k=1}^{N-1} f\left(a+k\frac{b-a}{N}\right)\right] - \int_a^b f(x)dx, \ \sum_{k=1}^{N} \frac{f''(\xi)h^3}{12} = \frac{f''(\xi)h^3 N}{12}$$

So that:

$$-\frac{f''(\xi)h^3 N}{12} \le \frac{b-a}{N}\left[\frac{f(a)+f(b)}{2} + \sum_{k=1}^{N-1} f\left(a+k\frac{b-a}{N}\right)\right] - \int_a^b f(x)dx \le \frac{f''(\xi)h^3 N}{12}$$

$$\text{error} = \int_a^b f(x)dx - \frac{b-a}{N}\left[\frac{f(a)+f(b)}{2} + \sum_{k=1}^{N-1} f\left(a+k\frac{b-a}{N}\right)\right] = \frac{f''(\xi)h^3 N}{12} = \frac{f''(\xi)(b-a)^3}{12N^2}$$

In general, three techniques are used in the analysis of error:

1. Fourier series

2. Residue calculus

3. Euler–Maclaurin summation formula:

   An asymptotic error estimate for $N \to \infty$ is given by,

$$\text{error} = -\frac{(b-a)^2}{12N^2}\left[f'(b)-f'(a)\right]+O(N^{-3}).$$

Further terms in this error estimate are given by the Euler–Maclaurin summation formula.

It is argued that the speed of convergence of the trapezoidal rule reflects and can be used as a definition of classes of smoothness of the functions.

Periodic and Peak Functions

The trapezoidal rule converges rapidly for periodic functions. This is an easy consequence of the Euler-Maclaurin summation formula, which says that if f is p times continuously differentiable with period,

$$\sum_{k=0}^{N-1} f(kh)h = \int_0^T f(x)dx + \sum_{k=1}^{\lfloor p/2 \rfloor} \frac{B_{2k}}{(2k)!}(f^{(2k-1)}(T)-f^{(2k-1)}(0))-(-1)^m h^m \int_0^T \tilde{B}_p(x/T)f^{(p)}(x)dx$$

where $h := T/N$ and $\tilde{B}_p$ is the periodic extension of the pth Bernoulli polynomial. Due to the periodicity, the derivatives at the endpoint cancel and we see that the error is $O(h^m)$.

A similar effect is available for peak-like functions, such as Gaussian, Exponentially modified Gaussian and other functions with derivatives at integration limits that can be neglected. The evaluation of the full integral of a Gaussian function by trapezoidal rule with 1% accuracy can be made using just 4 points. Simpson's rule requires 1.8 times more points to achieve the same accuracy.

Although some effort has been made to extend the Euler-Maclaurin summation formula to higher dimensions, the most straightforward proof of the rapid convergence of the trapezoidal rule in higher dimensions is to reduce the problem to that of convergence of Fourier series. This line of reasoning shows that if f is periodic on a n-dimensional space with p continuous derivatives, the speed of convergence is $O(h^{p/d})$. For very large dimension, the shows that Monte-Carlo integration is most likely a better choice, but for 2 and 3 dimensions, equispaced sampling is efficient. This is exploited in computational solid state physics where equispaced sampling over primitive cells in the reciprocal lattice is known as *Monkhorst-Pack integration*.

Example:

The vertical distance covered by a rocket from $t = 8$ to $t = 30$ seconds is given by,

$$x = \int_8^{30}\left(2000\ln\left[\frac{140000}{140000-2100t}\right]-9.8t\right)dt$$

a. Use the single segment trapezoidal rule to find the distance covered for $t = 8$ to $t = 30$ seconds.

b. Find the true error, E_t for part (a).

c. Find the absolute relative true error for part (a).

Solution

a) $I \approx (b-a) \left[\dfrac{f(a) + f(b)}{2} \right]$, where

$a = 8$

$b = 30$

$$f(t) = 2000 \ln \left[\frac{140000}{140000 - 2100t} \right] - 9.8t$$

$$f(8) = 2000 \ln \left[\frac{140000}{140000 - 2100(8)} \right] - 9.8(8)$$

$= 177.27$ m/s

$$f(30) = 2000 \ln \left[\frac{140000}{140000 - 2100(30)} \right] - 9.8(30)$$

$= 901.67$ m/s

$$I \approx (30 - 8) \left[\frac{177.27 + 901.67}{2} \right]$$

$= 11868$ m

b) The exact value of the above integral is,

$$x = \int_{8}^{30} \left(2000 \ln \left[\frac{140000}{140000 - 2100t} \right] - 9.8t \right) dt$$

$= 11061$ m

so the true error is,

E_t = True Value – Approximate Value

$= 11061 - 11868$

$= -807$ m

c) The absolute relative true error $|\epsilon_t|$, would then be,

$$|\epsilon_t| = \left|\frac{\text{True Error}}{\text{True Value}}\right| \times 100$$

$$= \left|\frac{11061-11868}{11061}\right| \times 100$$

$$= 7.2958\%$$

Multiple-segment Trapezoidal Rule

The true error using a single segment trapezoidal rule was large. We can divide the interval [8,30] into [8,19] and [19,30] intervals and apply the trapezoidal rule over each segment.

$$f(t) = 2000\ln\left(\frac{140000}{140000-2100t}\right) - 9.8t$$

$$\approx (19-8)\left[\frac{f(8)+f(19)}{2}\right] + (30-19)\left[\frac{f(19)+f(30)}{2}\right]$$

$f(8) = 177.27$ m/s

$$f(19) = 2000\ln\left[\frac{140000}{140000-2100(19)}\right] - 9.8(19) = 484.75 \text{ m/s}$$

$f(30) = 901.67$ m/s

Hence

$$\int_8^{30} f(t)dt \approx (19-8)\left[\frac{177.27+484.75}{2}\right] + (30-19)\left[\frac{484.75+901.67}{2}\right]$$

$$= 11266 \text{ m}$$

The true error, E_t is,

$$E_t = 11061-11266$$

$$= -205\text{m}$$

The true error now is reduced from 807 m to 205 m. Extending this procedure to dividing [a,b] into n equal segments and applying the trapezoidal rule over each segment, the sum of the results obtained for each segment is the approximate value of the integral. Divide (b − a) into n equal segments as shown in figure below. Then the width of each segment is,

$$h = \frac{b-a}{n}$$

The integral I can be broken into h integrals as,

$$I = \int_a^b f(t)dx$$

$$= \int_a^{a+h} f(x)dx + \int_{a+h}^{a+2h} f(x)dx + \ldots + \int_{a+(n-2)h}^{a+(n-1)h} f(x)dx + \int_{a+(n-1)h}^{b} f(x)dx$$

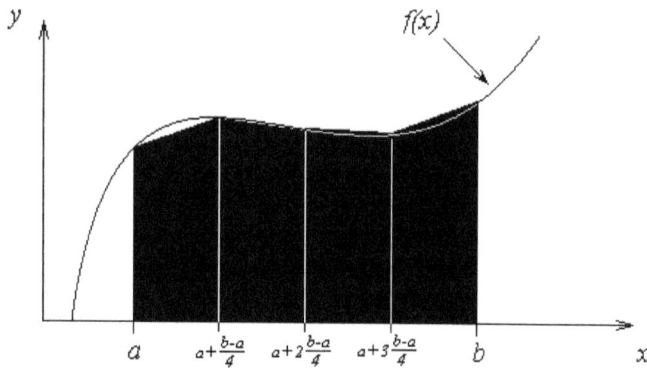

Figure: Multiple ($n = 4$) segment trapezoidal rule

Applying trapezoidal rule above equation on each segment gives,

$$\int_a^b f(x)dx = [(a+h)-a]\left[\frac{f(a)+f(a+h)}{2}\right]$$

$$+[(a+2h)-(a+h)]\left[\frac{f(a+h)+f(a+2h)}{2}\right]$$

$$+\ldots\ldots\ldots+[(a+(n-1)h)-(a+(n-2)h)]\left[\frac{f(a+(n-2)h)+f(a+(n-1)h)}{2}\right]$$

$$+[b-(a+(n-1)h)]\left[\frac{f(a+(n-1)h)+f(b)}{2}\right]$$

$$+h\left[\frac{f(a+(n-2)h)+f(a+(n-1)h)}{2}\right]+h\left[\frac{f(a+(n-1)h)+f(b)}{2}\right]$$

$$=h\left[\frac{f(a)+2f(a+h)+2f(a+2h)+\ldots+2f(a+(n-1)h)+f(b)}{2}\right]$$

$$= \frac{h}{2}\left[f(a) + 2\left\{\sum_{i=1}^{n-1} f(a+ih)\right\} + f(b)\right]$$

$$= \frac{b-2}{2n}\left[f(a) + 2\left\{\sum_{i=1}^{n-1} f(a+ih)\right\} + f(b)\right]$$

Example:

The vertical distance covered by a rocket from $t = 8$ to $t = 30$ seconds is given by

$$x = \int_{8}^{30}\left(2000\ln\left[\frac{140000}{140000 - 2100t}\right] - 9.8t \right) dt$$

a) Use the two-segment trapezoidal rule to find the distance covered from $t = 8$ to $t = 30$ seconds.

b) Find the true error, E_t for part (a).

c) Find the absolute relative true error for part (a).

Solution

a) The solution using 2-segment Trapezoidal rule is

$$I \approx \frac{b-2}{2n}\left[f(a) + 2\left\{\sum_{i=1}^{n-1} f(a+ih)\right\} + f(b)\right]$$

n = 2

a = 8

b = 30

$$h = \frac{b-a}{n}$$

= 11

$$I \approx \frac{30-8}{2(2)}\left[f(8) + 2\left\{\sum_{i=1}^{2-1} f(8+11i)\right\} + f(30)\right]$$

$$= \frac{22}{4}[f(8) + 2f(19) + f(30)]$$

$$= \frac{22}{4}[177.27 + 2(484.75) + 9.1.67]$$

$$= 11266 \text{ m}$$

b) The exact value of the above integral is,

$$x = \int_{8}^{30} \left(2000 \ln\left[\frac{140000}{140000 - 2100t} \right] - 9.8t \right) dt$$

= 11061 m

so the true error is,

E_t = True Value – Approximate Value

= 11061–11266

= –205 m

c) The absolute relative true error, $|\epsilon_t|$, would then be,

$$|\epsilon_t| = \left| \frac{\text{True Error}}{\text{True Value}} \right| \times 100$$

$$= \left| \frac{11061 - 11266}{11061} \right| \times 100$$

$$= 1.8537\%$$

Table: Values obtained using multiple-segment trapezoidal rule for,

$$x = \int_{8}^{30} \left(2000 \ln\left[\frac{140000}{140000 - 2100t} \right] - 9.8t \right) dt$$

n	Approximate Value	Et	$|\epsilon_t|$%	$|\epsilon_a|$%
1	11868	-807	7.296	---
2	11266	-205	1.853	5.343
3	11153	-91.4	0.8265	1.019
4	11113	-51.5	0.4655	0.3594
5	11094	-33.0	0.2981	0.1669
6	11084	-22.9	0.2070	0.09082
7	11078	-16.8	0.1521	0.05482
8	11074	-12.9	0.1165	0.03560

Example:

Use the multiple-segment trapezoidal rule to find the area under the curve

$$f(x) = \frac{300x}{1 + e^x}$$

from $x = 0$ to $x = 10$.

Solution

Using two segments, we get,

$$h = \frac{10-0}{2} = 5$$

$$f(0) = \frac{300(0)}{1+e^0} = 0$$

$$f(5) = \frac{300(5)}{1+e^5} = 10.039$$

$$f(10) = \frac{300(10)}{1+e^{10}} = 0.136$$

$$I \approx \frac{b-2}{2n}\left[f(a) + 2\left\{\sum_{i=1}^{n-1}f(a+ih)\right\} + f(b)\right]$$

$$= \frac{10-0}{2(2)}\left[f(0) + 2\left\{\sum_{i=1}^{2-1}f(0+5)\right\} + f(10)\right]$$

$$= \frac{10}{4}\left[f(0) + 2f(5) + f(10)\right]$$

$$= \frac{10}{4}\left[0 + 2(10.039) + 0.136\right] = 50.537$$

So what is the true value of this integral?

$$\int_0^{10} \frac{300x}{1+e^x}dx = 246.59$$

Making the absolute relative true error,

$$|\epsilon_t| = \left|\frac{246.59 - 50.535}{246.59}\right| \times 100$$

$$= 79.506\%$$

Why is the true value so far away from the approximate values? Just take a look at figure below. As you can see, the area under the "trapezoids" covers a small portion of the area under the curve. As we add more segments, the approximated value quickly approaches the true value.

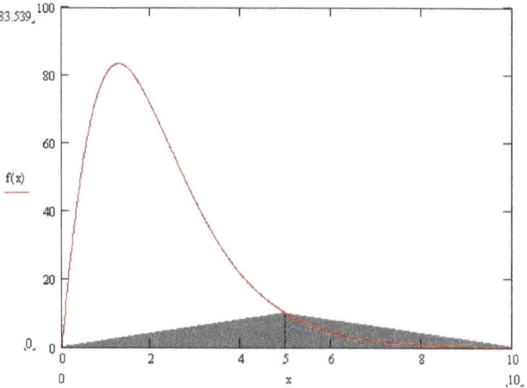

Figure: 2-segment trapezoidal rule approximation

Table: Values obtained using multiple-segment trapezoidal rule for $\int_{0}^{10} \dfrac{300x}{1+e^x} dx$.

n	Approximate Value	E_t	$|\in_t|\%$
1	0.681	245.91	99.724%
2	50.535	196.05	79.505%
4	170.61	75.978	30.812%
8	227.04	19.546	7.927%
16	241.70	4.887	1.982%
32	245.37	1.222	0.495%
64	246.28	0.305	0.124%

Example:

Use multiple-segment trapezoidal rule to find

$$I = \int_{0}^{2} \frac{1}{\sqrt{x}} dx$$

Solution

We cannot use the trapezoidal rule for this integral, as the value of the integrand at $x = 0$ is infinite. However, it is known that a discontinuity in a curve will not change the area under it. We can assume any value for the function at $x = 0$. The algorithm to define the function so that we can use the multiple-segment trapezoidal rule is given below.

Function $f(x)$

If $x = 0$ Then $f = 0$

If $x \neq 0$ Then $f = x^{\wedge} (-0.5)$

End Function

Basically, we are just assigning the function a value of zero at $x = 0$. Everywhere else, the function is continuous. This means the true value of our integral will be just that—true. Let's see what happens using the multiple-segment trapezoidal rule. Using two segments, we get,

$$h = \frac{2-0}{2} = 1$$

$$f(0) = 0$$

$$f(1) = \frac{1}{\sqrt{1}} = 1$$

$$f(2) = \frac{1}{\sqrt{2}} = 0.70711$$

$$I \approx \frac{b-2}{2n} \left[f(a) + 2 \left\{ \sum_{i=1}^{n-1} f(a+ih) \right\} + f(b) \right]$$

$$= \frac{2-0}{2(2)} \left[f(0) + 2 \left\{ \sum_{i=1}^{2-1} f(0+1) \right\} + f(2) \right]$$

$$= \frac{2}{4} \left[f(0) + 2f(1) + f(2) \right]$$

$$= \frac{2}{4} \left[0 + 2(1) + 0.70711 \right]$$

$$= 1.3536$$

So what is the true value of this integral?

$$\int_0^2 \frac{1}{\sqrt{x}} dx = 2.8284$$

Thus making the absolute relative true error,

$$|\epsilon_t| = \left| \frac{2.8284 - 1.3536}{2.8284} \right| \times 100$$

$$= 52.145\%$$

Table: Values obtained using multiple-segment trapezoidal rule for $\int_{0}^{2} \frac{1}{\sqrt{x}} dx$.

n	Approximate Value	E_t	$|\epsilon_t|$
2	1.354	1.474	52.14%
4	1.792	1.036	36.64%
8	2.097	0.731	25.85%
16	2.312	0.516	18.26%
32	2.463	0.365	12.91%
64	2.570	0.258	9.128%
128	2.646	0.182	6.454%
256	2.699	0.129	4.564%
512	2.737	0.091	3.227%
1024	2.764	0.064	2.282%
2048	2.783	0.045	1.613%
4096	2.796	0.032	1.141%

Error in Multiple-segment Trapezoidal Rule

The true error for a single segment Trapezoidal rule is given by,

$$E_t = -\frac{(b-a)^3}{12} f''(\zeta), a < \zeta < b$$

Where ζ is some point in [a,b].

What is the error then in the multiple-segment trapezoidal rule? It will be simply the sum of the errors from each segment, where the error in each segment is that of the single segment trapezoidal rule. The error in each segment is,

$$E_1 = -\frac{[(a+h)-a]^3}{12} f''(\zeta_1), a < \zeta_1 < a+h$$

$$= -\frac{h^3}{12} f''(\zeta_1)$$

$$E_2 = -\frac{[(a+2h)-(a+h)]^3}{12} f''(\zeta_2), a+h < \zeta_2 < a+2h$$

$$= -\frac{h^3}{12} f''(\zeta_2)$$

$$E_i = -\frac{[(a+ih)-(a+(i-1)h)]^3}{12} f''(\zeta_i), a+(i-1)h < \zeta_i < a+ih$$

$$= -\frac{h^3}{12} f''(\zeta_i)$$

$$E_{n-1} = -\frac{[\{a+(n-1)h\}-\{a+(n-2)h\}]^3}{12} f''(\zeta_{n-1}), a+(n-2)h < \zeta_{n-1} < a+(n-1)h$$

$$= -\frac{h^3}{12} f''(\zeta_{n-1})$$

$$E_n = -\frac{[b-\{a+(n-1)h\}]^3}{12} f''(\zeta_n), a+(n-1)h < \zeta_n < b$$

$$= -\frac{h^3}{12} f''(\zeta_n)$$

Hence the total error in the multiple-segment trapezoidal rule is,

$$E_t = \sum_{i=1}^{n} E_i$$

$$= -\frac{h^3}{12} \sum_{i=1}^{n} f''(\zeta_i)$$

$$= -\frac{(b-a)^3}{12n^3} \sum_{i=1}^{n} f''(\zeta_i)$$

$$= -\frac{(b-a)^3}{12n^2} \sum_{i=1}^{n} f''(\zeta_i)$$

The term $\dfrac{\sum_{i=1}^{n} f''(\zeta_i)}{n}$ is an approximate average value of the second derivative $f''(x), a < x < b$.

Hence

$$E_t = -\frac{(b-a)^3}{12n^2} \frac{\sum_{i=1}^{n} f''(\zeta_i)}{n}$$

In table below, the approximate value of the integral,

$$\int_{8}^{30} \left(2000 \ln\left[\frac{140000}{140000-2100t} \right] - 9.8t \right) dt$$

is given as a function of the number of segments. You can visualize that as the number of segments are doubled, the true error gets approximately quartered.

Table: Values obtained using multiple-segment trapezoidal rule for

$$x = \int_{8}^{30} \left(2000 \ln \left[\frac{140000}{140000 - 2100t} \right] - 9.8t \right) dt.$$

n	Approximate Value	E_t	$\left| \in_t \right| \%$	$\left| \in_a \right| \%$
2	11266	-205	1.853	5.343
4	11113	-52	0.4701	0.3594
8	11074	-13	0.1175	0.03560
16	11065	-4	0.03616	0.00401

Simpson's Rule

Simpson's rule is a numerical method that approximates the value of a definite integral by using quadratic polynomials.

Let's first derive a formula for the area under a parabola of equation $y = ax^2 + bx + c$ passing through the three points: $(-h, y_0), (0, y_1), (h, y_2)$.

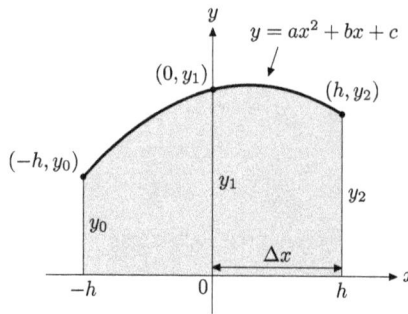

$$A = \int_{-h}^{h} (ax^2 + bx + c)dx$$

$$= \left(\frac{ax^3}{3} + \frac{bx^2}{2} + cx \right) \Bigg|_{-h}^{h}$$

$$= \frac{2ah^3}{3} + 2ch$$

$$= \frac{h}{3}(2ah^2 + 6c)$$

Since the point $(-h, y_0), (0, y_1), (h, y_2)$ are on the parabola, they satisfy $y = ax^2 + bx + c$. Therefore

$$y_0 = ah^2 - bh + c$$
$$y_1 = c$$
$$y_2 = ah^2 + bh + c$$

Observe that

$$y_0 + 4y_1 + y_2 = (ah^2 - bh + c) + 4c + (ah^2 + bh + c) = 2ah^2 + 6c.$$

Therefore, the area under the parabola is,

$$A = \frac{h}{3}(y_0 + 4y_1 + y_2) = \frac{\Delta x}{3}(y_0 + 4y_1 + y_2).$$

We consider the definite integral,

$$\int_a^b f(x)dx.$$

We assume that f(x) is continuous on [a, b] and we divide [a, b] into an even number n of subintervals of equal length,

$$\Delta x = \frac{b - a}{n}$$

using the n + 1 points,

$$x_0 = a, \quad x_1 = a + \Delta x, \quad x_2 = a + 2\Delta x, \ldots, \quad x_n = a + n\Delta x = b.$$

We can compute the value of f(x) at these points.

$$y_0 = f(x_0), \quad y_1 = f(x_1), \quad y_2 = f(x_2), \ldots, \quad y_n = f(x_n).$$

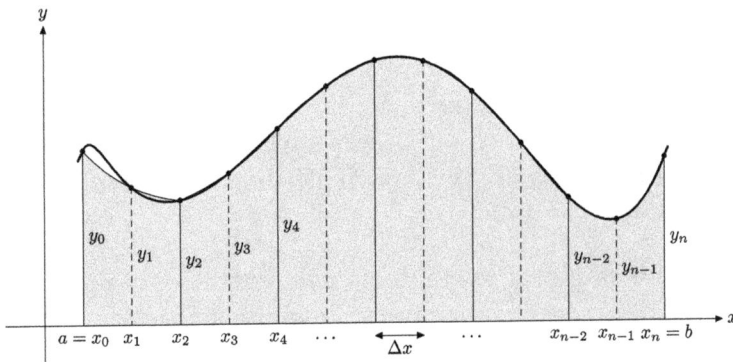

We can estimate the integral by adding the areas under the parabolic arcs through three successive points.

$$\int_a^b f(x)dx \approx \frac{\Delta x}{3}(y_0 + 4y_1 + y_2) + \frac{\Delta x}{3}(y_2 + 4y_3 + y_4) + \cdots + \frac{\Delta x}{3}(y_{n-2} + 4y_{n-1} + y_n)$$

By simplifying, we obtain Simpson's rule formula.

$$\int_a^b f(x)dx \approx \frac{\Delta x}{3}(y_0 + 4y_1 + 2y_2 + 4y_3 + 2y_4 + \cdots + 4y_{n-1} + y_n)$$

Example: Use Simpson's rule with n = 6 to estimate

$$\int_1^4 \sqrt{1 + x^3}\, dx.$$

Solution: For n = 6, we have $\Delta x = \dfrac{4-1}{6} = 0.5$. We compute the values of $y_0, y_1, y_2, \ldots, y_6$.

X	1	1.5	2	2.5	3	3.5	4
$y = \sqrt{1 + x^3}$	$\sqrt{2}$	$\sqrt{4.375}$	3	$\sqrt{16.625}$	$\sqrt{28}$	$\sqrt{43.875}$	$\sqrt{65}$

Therefore,

$$\int_1^4 \sqrt{1 + x^3}\, dx \approx \frac{0.5}{3}(\sqrt{2} + 4\sqrt{4.375} + 2(3) + 4\sqrt{16.625} + 2\sqrt{28} + 4\sqrt{43.875} + \sqrt{65})$$

$$\approx 12.871$$

Error

The error in approximating an integral by Simpson's rule is

$$-\frac{1}{90}\left(\frac{b-a}{2}\right)^5 f^{(4)}(\xi),$$

where ξ is some number between a and b.

The error is asymptotically proportional to $(b-a)^5$. However, the above derivations suggest an error proportional to $(b-a)^4$. Simpson's rule gains an extra order because the points at which the integrand is evaluated are distributed symmetrically in the interval $[a, b]$.

Since the error term is proportional to the fourth derivative of f at ξ, this shows that Simpson's rule provides exact results for any polynomial f of degree three or less, since the fourth derivative of such a polynomial is zero at all points.

Simpson's 3/8 rule

Simpson's 3/8 rule is another method for numerical integration proposed by Thomas Simpson. It is based upon a cubic interpolation rather than a quadratic interpolation. Simpson's 3/8 rule is as follows:

$$\int_a^b f(x)dx \approx \frac{3h}{8}\left[f(a)+3f\left(\tfrac{2a+b}{3}\right)+3f\left(\tfrac{a+2b}{3}\right)+f(b)\right]=\frac{(b-a)}{8}\left[f(a)+3f\left(\tfrac{2a+b}{3}\right)+3f\left(\tfrac{a+2b}{3}\right)+f(b)\right],$$

where $b - a = 3h$. The error of this method is:

$$-\frac{(b-a)^5}{6480}f^{(4)}(\xi),$$

where ξ is some number between a and b. Thus, the 3/8 rule is about twice as accurate as the standard method, but it uses one more function value. A composite 3/8 rule also exists, similarly as above.

A further generalization of this concept for interpolation with arbitrary-degree polynomials are the Newton–Cotes formulas.

Composite Simpson's 3/8 Rule

Dividing the interval $[a,b]$ into n subintervals of length $h=(b-a)/n$ and introducing the nodes $x_i = a+ih$ we have,

$$\int_a^b f(x)dx \approx \frac{3h}{8}\left[f(x_0)+3f(x_1)+3f(x_2)+2f(x_3)+3f(x_4)+3f(x_5)+2f(x_6)+\cdots+f(x_n)\right]$$

$$=\frac{3h}{8}\left[f(x_0)+3\sum_{i\neq 3k}^{n-1}f(x_i)+2\sum_{j=1}^{n/3-1}f(x_{3j})+f(x_n)\right] \quad \text{For: } k\in\mathbb{N}_0$$

While the remainder for the rule is shown as:

$$-\frac{h^4}{80}(b-a)f^{(4)}(\xi),$$

Note, we can only use this if n is a multiple of three.

A simplified version of Simpson's rules is used in naval architecture. The 3/8th rule is also called Simpson's second rule.

Alternative Extended Simpson's Rule

This is another formulation of a composite Simpson's rule: instead of applying Simpson's rule to disjoint segments of the integral to be approximated, Simpson's rule is applied to overlapping segments, yielding:

$$\int_a^b f(x)dx \approx \tfrac{h}{48}\Big[17f(x_0)+59f(x_1)+43f(x_2)+49f(x_3)+48\sum_{i=4}^{n-4}f(x_i)+49f(x_{n-3})+43f(x_{n-2})+59f(x_{n-1})+17f(x_n)\Big].$$

The formula above is obtained by combining the original composite Simpson's rule with the one consisting of using Simpson's 3/8 rule in the extreme subintervals and the standard 3-point rule in the remaining subintervals. The result is then obtained by taking the mean of the two formulas.

Simpson's Rules in the Case of Narrow Peaks

In the task of estimation of full area of narrow peak-like functions, Simpson's rules are much less efficient than trapezoidal rule. Namely, composite Simpson's 1/3 rule requires 1.8 times more points to achieve the same accuracy as trapezoidal rule. Composite Simpson's 3/8 rule is even less accurate. Integral by Simpson's 1/3 rule can be represented as a sum of 2/3 of integral by trapezoidal rule with step h and 1/3 of integral by rectangle rule with step 2h. No wonder that error of the sum corresponds lo less accurate term. Averaging of Simpson's 1/3 rule composite sums with properly shifted frames produces following rules:

$$\int_a^b f(x)dx \approx \tfrac{h}{24}\Big[-f(x_0)+12f(x_1)+23f(x_2)+24\sum_{i=2}^{n-2}f(x_i)+23f(x_{n-1})+12f(x_n)-f(x_{n+1})\Big]$$

where two points outside of integrated region are exploited and

$$\int_a^b f(x)dx \approx \tfrac{h}{24}\Big[9f(x_1)+28f(x_2)+23f(x_3)+24\sum_{i=3}^{n-3}f(x_i)+23f(x_{n-2})+28f(x_{n-1})+9f(x_n)\Big]$$

Those rules are very much similar to Press's alternative extended Simpson's rule. Coefficients within the major part of the region being integrated equal one, differences are only at the edges. These three rules can be associated with Euler-MacLaurin formula with the first derivative term and named Euler-MacLaurin integration rules. They differ only in the way, how the first derivative at the region end is calculated.

Newton–Cotes Formulas

The Newton-Cotes formulas is the most commonly used numerical integration methods to approximate the integration of a complicated function by replacing the function with many polynomials across the integration interval. The integration of the original function can then be obtained by summing up all polynomials whose "areas" are calculated by the weighting coefficients and the values of the function at the nodal points.

Closed Newton-Cotes Formulas

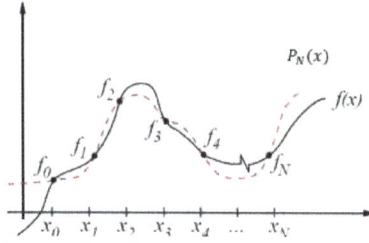

Figure: Closed Newton-Cotes Formulas

Let $a = x_0$; $b = x_N$; and $h = \dfrac{b-a}{N}$. $x_i = x_0 + ih$, for $i = 0, 1, \ldots, N$.

$$\int_a^b f(x)dx \approx \sum_{i=0}^{N} a_i f(x_i) \quad \text{with } a_i = \int_a^b L_{N,i}(x)dx.$$

Here $L_{N,i}(x)$ is the ith Lagrange base polynomial of degree N.

Theorem: Suppose that $\sum_{i=0}^{N} a_i f(x_i)$ is the (n+1)-point closed Newton-Cotes formula with $a = x_0$; $b = x_N$; and $h = \dfrac{b-a}{N}$. There exists $\xi \in (a,b)$ for which

$$\int_a^b f(x)dx \approx \sum_{i=0}^{N} a_i f(x_i) + \frac{h^{N+3} f^{(N+2)}(\xi)}{(N+2)!} \int_0^N t^2(t-1)\cdots(t-N)dt, \quad \text{if } N \text{ is even and}$$

$f \in C^{N+2}[a,b]$, and

$$\int_a^b f(x)dx \approx \sum_{i=0}^{N} a_i f(x_i) + \frac{h^{N+2} f^{(N+1)}(\xi)}{(N+1)!} \int_0^N t^2(t-1)\cdots(t-N)dt$$

if N is odd and $f \in C^{N+1}[a,b]$.

Remark: N is even, degree of precision is $N+1$. N is odd, degree of precision is N.

Examples: N=1: Trapezoidal rule; N=2: Simpson's rule.

N=3: Simpson's Three-Eighths rule

$$\int_{x_0}^{x_3} f(x)dx = \frac{3h}{8}(f(x_0) + 3f(x_1) + 3f(x_2) + f(x_3)) - \frac{3h^5}{80} f^{(4)}(\xi) \quad \text{where } x_0 < \xi < x_3; h = \frac{x_3 \quad x_0}{}.$$

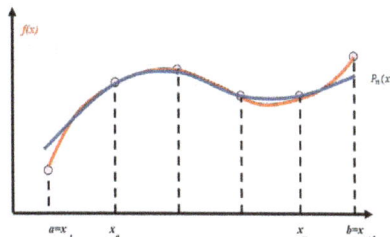

Figure: Open Newton-Cotes Formula

Open Newton-Cotes Formula

Let $h = \dfrac{b-a}{n+2}$; and $x_0 = a + h$. $x_1 = x_0 + ih$, for $i = 0, 1, \ldots, n$. This implies $x_n = b - h$.

Theorem: Suppose that $\sum_{i=0}^{N} a_i f(x_i)$ is the (n+1)-point open Newton-Cotes formula with $a = x_{-1}; b = x_{n+1};$ and $h = \dfrac{b-a}{n+2}$. There exists $\xi \in (a, b)$ for which

$$\int_a^b f(x)dx \approx \sum_{i=0}^{n} a_i f(x_i) + \frac{h^{n+3} f^{(n+2)}(\xi)}{(n+2)!} \int_{-1}^{n+1} t^2(t-1)\cdots(t-n)dt,$$

if is n even and $f \in C^{n+2}[a, b]$, and

$$\int_a^b f(x)dx \approx \sum_{i=0}^{n} a_i f(x_i) + \frac{h^{n+2} f^{(n+1)}(\xi)}{(n+1)!} \int_{-1}^{n+1} t^2(t-1)\cdots(t-n)dt,$$

If n is odd and $f \in C^{n+1}[a, b]$.

Examples: of open Newton-Cotes formulas,

n=0: Midpoint rule

$$\int_{x_{-1}}^{x_1} f(x)dx = 2hf(x_0) + \frac{h^3}{3} f^{(2)}(\xi)$$

Where $x_{-1} < \xi < x_1$. $h = \dfrac{b-a}{2}$

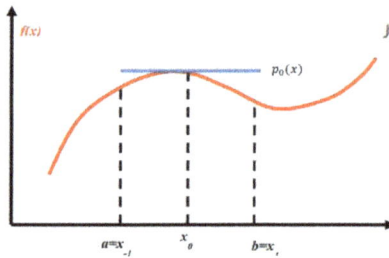

Figure: Midpoint rule

$$n = 1: \int_{x_{-1}}^{x_2} f(x)dx = \frac{3h}{2}[f(x_0) + f(x_1)] + \frac{3h^3}{4} f^{(2)}(\xi) \quad \text{where } x_{-1} < \xi < x_2. \, h = \frac{b-a}{3}$$

$$n = 2: \int_{x_{-1}}^{x_3} f(x)dx = \frac{4h}{3}[2f(x_0) + f(x_2)] + \frac{14h^5}{45} f^{(4)}(\xi) \quad \text{where } x_{-1} < \xi < x_3. \, h = \frac{b-a}{4}$$

Closed Newton–Cotes formulas

This table lists some of the Newton–Cotes formulas of the closed type. For $0 \le i \le n$, with n the degree, let $x_i = a + i\frac{b-a}{n} = a + ih,$ and the notation f_i be a shorthand for $f(x_i)$.

Closed Newton–Cotes Formulas				
Degree n	Step size h	Common names	Formula	Error term
1	$h = (b-a)$	Trapezoid rule	$\dfrac{h}{2}(f_0 + f_1)$	$-\dfrac{(b-a)^3}{12}f^{(2)}(\xi)$
2	$h = \dfrac{(b-a)}{2}$	Simpson's rule	$\dfrac{h}{3}(f_0 + 4f_1 + f_2)$	$-\dfrac{(b-a)^5}{90 \cdot 2^5}f^{(4)}(\xi)$
3	$h = \dfrac{(b-a)}{3}$	Simpson's 3/8 rule	$3/8 \cdot h(f_0 + 3f_1 + 3f_2 + f_3)$	$-\dfrac{3(b-a)^5}{80 \cdot 3^5}f^{(4)}(\xi)$
4	$h = \dfrac{(b-a)}{4}$	Boole's rule	$\dfrac{2h}{45}(7f_0 + 32f_1 + 12f_2 + 32f_3 + 7f_4)$	$-\dfrac{8(b-a)^7}{945 \cdot 4^7}f^{(6)}(\xi)$

Boole's rule is sometimes mistakenly called Bode's rule, as a result of the propagation of a typographical error in Abramowitz and Stegun, an early reference book.

The exponent of the segment size $b - a$ in the error term shows the rate at which the approximation error decreases. The degree of the derivative of f in the error term gives the degree up to which polynomials can be integrated exactly (i.e., with error equal to zero) with this rule. Note that the derivative of f in the error term increases by 2 for every other rule. The number ξ must be taken from the interval (a,b).

Open Newton–Cotes Formulas

This table lists some of the Newton–Cotes formulas of the open type. Again, f_i is a shorthand for $f(x_i)$, with $x_i = a + i(b - a)/n$, and n the degree.

Open Newton–Cotes Formulas				
Common name	step size h	Formula	Error term	Degree
Rectangle rule, or midpoint rule	$\dfrac{b-a}{2}$	$2hf_1$	$\dfrac{1}{3}h^3 f^{(2)}(\xi)$	2
Trapezoid method	$\dfrac{b-a}{3}$	$\dfrac{3}{2}h(f_1 + f_2)$	$\dfrac{1}{4}h^3 f^{(2)}(\xi)$	3
Milne's rule	$\dfrac{b-a}{4}$	$\dfrac{4}{3}h(2f_1 - f_2 + 2f_3)$	$\dfrac{28}{90}h^5 f^{(4)}(\xi)$	4
No name	$\dfrac{b-a}{5}$	$\dfrac{5}{24}h(11f_1 + f_2 + f_3 + 11f_4)$	$\dfrac{95}{144}h^5 f^{(4)}(\xi)$	5

Instability for High Degree

A Newton–Cotes formula of any degree n can be constructed. However, for large n a Newton–Cotes rule can sometimes suffer from catastrophic Runge's phenomenon where the error grows exponentially for large n. Methods such as Gaussian quadrature and Clenshaw–Curtis quadrature with unequally spaced points (clustered at the *endpoints* of the integration interval) are stable and much more accurate, and are normally preferred to Newton–Cotes. If these methods cannot be used, because the integrand is only given at the fixed equidistributed grid, then Runge's phenomenon can be avoided by using a composite rule.

Alternatively, stable Newton–Cotes formulas can be constructed using least-squares approximation instead of interpolation. This allows building numerically stable formulas even for high degrees.

Gaussian Quadrature

The rule to approximate the definite integral value of a function is named as Gaussian quadrature rule. The weighted sum of the function values at particular points in certain domain (limits) of integration is known as Gaussian quadrature.

This formula is applicable only when the function f (x) is explicitly known in terms of its argument and the limits of integration are -1 to +1. According to this formula the integral is expressed as a weighted sum of n- terms in the form

$$\int_{-1}^{1} f(x)dx = \sum_{i=1}^{n} W_i \, f(x_i)$$

where $w_1, w_2, \ldots, w_n$ are unknown coefficients known as weights and $-1 < x_1 < x_2 < \cdots < x_n < 1$ are also unknown values of x called weighting factors. This is known as n-terms formula in which there are 2n unknowns. These 2n unknowns will be evaluated by requiring that above formula be true for all polynomials of degree (2n-1) { i.e. one less to the number of unknowns} or less. Then, it must certainly be true for simplest cases i.e. for $x^0, x^1, x^2, \ldots x^{2n-1}$,. The formulae based upon this principle are known as Gaussian quadrature formulae. The satisfaction of above formula for $f(x)x^0, x^1, x^2, \ldots x^{2n-1}$ will lead to a set of 2n linear equations in $w_i's$. Their solution will give the values of $x_i's$ and $w_i's$. Substituting back these values in above formula, one gets n-terms Gaussian quadrature formula.

Example: Derivation of 2-terms formula:

For 2-terms formula, above formula becomes,

$$\int_{-1}^{+1} f(x)dx = w_1 f(x_1) + w_2 f(x_2)$$

Since above formula has four unknowns, as per principle of the method, it must be true for all polynomials of degree 3 or less. Then, it must certainly be true for the simplest cases $f(x)x^0, x^1, x^2, x^3$

$$f(x) = x^0; \int_{-1}^{1} x^0 dx \equiv 2 = w_1.1 + w_2.1$$

$$f(x) = x^1; \int_{-1}^{1} x^1 dx = 0 = w_1 x_1 + w_2 x_2$$

$$f(x) = x^2; \int_{-1}^{1} x^2 dx \equiv \frac{2}{3} = w_1 x_1^2 + w_2 x_2^2$$

$$f(x) = x^3; \int_{-1}^{1} x^3 dx = 0 = w_1 x_1^3 + w_2 x_2^3$$

From equations $f(x) = x^1; \int_{-1}^{1} x^1 dx = 0 = w_1 x_1 + w_2 x_2$

and ($f(x) = x^3; \int_{-1}^{1} x^3 dx = 0 = w_1 x_1^3 + w_2 x_2^3$) i.e. (2)× x_1^2 −

$f(x) = x^3; \int_{-1}^{1} x^3 dx = 0 = w_1 x_1^3 + w_2 x_2^3$), we get $w_2 x_2 (x_1^2 - x_2^2) = 0 \Rightarrow$ either

$w_2 = 0$ or $x_2 = 0$ or $x_1 = \pm x_2$.

The only valid solution is $x_2 = -x_1$ other reduces our formula (D) into a single term or invalid.

We then find $w_1 = w_2 = 1$ and $-x_1 = x_2 = \dfrac{1}{\sqrt{3}}$.

Thus, two-term formula becomes,

$$\int_{-1}^{1} f(x) dx = f(-\frac{1}{\sqrt{3}}) + f(\frac{1}{\sqrt{3}}).$$

Similarly, the three-terms formula is given by,

$$\int_{-1}^{1} f(x) dx = w_1 f(x_1) + w_2 f(x_2) + w_3 f(x_3)$$

Where $w_1 = w_2 = \dfrac{5}{9}$, $w_2 = \dfrac{8}{9}$, $-x_1 = x_3 = \sqrt{3/5}$, $x_2 = 2$ and thus

$$\int_{-1}^{1} f(x) dx = \frac{5}{9} f\left(-\sqrt{\frac{3}{5}}\right) + \frac{8}{9} f(0) + \frac{5}{9} f(\sqrt{\frac{3}{5}})$$

In fact, the values of x in two-terms or three-terms formula are the zeros of Legendre polynomials $L_2(x), L_3(x)$, respectively.

Similarly, the higher terms formula can be obtained.

If the limits of integration are a to b instead of - 1 to +1, then the liner transformation,

$$x = \frac{(b-a)+(b+a)}{2} \text{ or } dx = (\frac{b-a}{2})dt$$

will change the limits as follows: $\int_a^b f(x)dx = (\frac{b-a}{2})\int_{-1}^1 f\left(\frac{(b-a)t+(b+a)}{2}\right)dt$

Example:

Calculate the value of the integral

$\int_4^{5.2} \log_e x \, dx$ by (a) Trapezoidal rule (b) Simpson's $\frac{1}{3}rd$ rule (c) Simpson's $\frac{3}{8}th$ rule

taking h = 0.2 retaining the numerical values to 7 decimals. Compare the results with exact values.

Solution:

x:	4.0	4.2	4.4	4.6	4.8	5.0	5.2
loge x :	1.3862944	1.4350845	1.4816045	1.5260563	1.5686159	1.6094379	1.6486586

(a) By Trapezoidal rule, we have,

$$\int_{4.0}^{5.2} \log_e x \, dx = \frac{h}{2}[y_{4.0} + 2\{y_{4.2} + y_{4.4} + y_{4.6} + y_{4.8} + y_{5.0}\} + y_{5.2}]$$
$$= 1.8276551$$

(b) By Simpson's $\frac{1}{3}rd$ rule, we have,

$$\int_{4.0}^{5.2} \log_e x \, dx = \frac{h}{3}[y_{4.0} + 4\{y_{4.2} + y_{4.6} + y_{5.0}\} + 2\{y_{4.4} + y_{4.8}\} + y_{5.2}]$$
$$= 1.8278472$$

(c) By Simpson's $\frac{3}{8}th$ rule, we have,

$$\int_{4.0}^{5.2} \log_e x \, dx = \frac{3}{8}h[y_{4.0} + 3\{y_{4.2} + y_{4.4} + y_{4.8} + y_{5.0}\} + 2 y_{4.6} + y_{5.2}]$$
$$= 1.8278470.$$

Actual value of $\int_{4.0}^{5.2} \log_e x \, dx = \left[x\{\log_e x - 1\}\right]_{4.0}^{5.2} = 1.8278475$

Hence the errors due to different formulae are,

(a) 0.0001924   (b) 0.0000003   (c) 0.0000005

We observe that Simpson's $\frac{1}{3}rd$ rule is more accurate.

Example:

Calculate by Simpson's $\frac{1}{3}rd$ rule an approximate value of the integral $\int_{-3}^{3} x^4 \, dx$ by taking seven equidistant ordinates. Compare it with the exact value and the value obtained by using the Trapezoidal rule.

Solution:

X	-3	-2	-1	0	1	2	3
y	81	16	1	0	1	16	81

By Simpson's rule, we get,

$$\int_{-3}^{3} x^4 dx = \frac{h}{3}\left[y_{-3} + 4\{y_{-2} + y_0 + y_2\} + 2\{y_{-2} + y_1\} + y_3 \right] = 98$$

The exact value of $\int_{-3}^{3} x^4 dx = \left(\frac{x^5}{5} \right)_{-3}^{3} = 97.2$

However, by Trapezoidal rule,

$$\int_{-3}^{3} x^4 dx = \frac{h}{2}\left[y_{-3} + 2\{y_{-2} + y_{-1} + y_0 + y_1 + y_2\} \, y_3 \right] = 115.$$

In this case we observe that the Trapezoidal rule does not give an accurate result.

Example:

Use Romberg's method to compute $I = \int_0^1 \frac{1}{1+x} dx$ correct to three decimal places with Trapezoidal rule.

Solution:

We take h = 0.5, 0.25 and 0.125 successively and using T.R., we get I(0.5) = 0.7084, (0.25) = 0.6970 and I (0.125) = 0.6941

Now, using the formula: $I_{improved} = I_{more\ correct} + \left\{ \dfrac{I_{more\ correct} = I_{less\ correct}}{2^n - 1} \right\}$

$I(0.5, 0.25) = 0.6970 + \dfrac{1}{3}(0.6970 - 0.7084) = 0.6932$

$I(0.25, 0.125) = 0.6941 + \dfrac{1}{3}(0.6941 - 0.6970) = 0.6931$

Finally,

$I(0.5, 0.25, 0.125) = 0.6931 \dfrac{1}{15}(0.6931 - 0.6932)$

= 0.6931

In tabular form

I	Error (h²)	Error (h4)	Error (h6)
0.5	0.7084	0.6932	0.6931
0.25	0.6970	0.6931	
0.125	0.6931		

Example:

Evaluate the integral

$I = \int_0^{\pi/2} \sin x \, dx$, using two-terms Gaussian formula

Solution: First we must change the limits of integration from $(0, \pi/2)$ to $(-1, 1)$.

Let $x = \dfrac{(\pi/2)t + \pi/2}{2} = \dfrac{\pi}{4}(t+1)$, so $dx = \dfrac{\pi}{4} dt$.

$$\therefore I = \int_{-1}^{1} \sin(\frac{\pi}{4}(t+1)) \left(\frac{\pi}{4}\right) dt$$

$$= \frac{\pi}{4} \left\{ 1.0 \cdot \sin\left(\frac{\pi}{4}\left\{1 - \frac{1}{\sqrt{3}}\right\}\right) + 1.0 \times \sin\left(\frac{\pi}{4}\left\{1 + \frac{1}{\sqrt{3}}\right\}\right) \right\}$$

$$= 0.99847$$

References

- Leader, Jeffery J. (2004). Numerical Analysis and Scientific Computation. Addison Wesley. ISBN 0-201-73499-0

- Weideman, J. A. C. (January 2002), "Numerical Integration of Periodic Functions: A Few Examples", The American Mathematical Monthly, 109 (1): 21–36, doi:10.2307/2695765, JSTOR 2695765

- Truncation-error: whatis.techtarget.com, Retrieved 19 May 2018

- Weisstein, Eric W. (2010). "Newton-Cotes Formulas". MathWorld--A Wolfram Web Resource. MathWorld. Retrieved 2 August 2010

- Cruz-Uribe, D.; Neugebauer, C. J. (2002), "Sharp Error Bounds for the Trapezoidal Rule and Simpson's Rule" (PDF), Journal of Inequalities in Pure and Applied Mathematics, 3 (4)

- Trapezoidal-rule-formula: byjus.com, Retrieved 29 June 2018

- Atkinson, Kendall E. (1989), An Introduction to Numerical Analysis (2nd ed.), New York: John Wiley & Sons, ISBN 978-0-471-50023-0

- Mathieu Ossendrijver (Jan 29, 2016). "Ancient Babylonian astronomers calculated Jupiter's position from the area under a time-velocity graph". Science. 351: 482–484. Bibcode:2016Sci...351..482O. doi:10.1126/science.aad8085. PMID 26823423

5

Numerical Differential Equations

It is not possible to find analytical solutions to many differential equations. Often, a numeric approximation to the solution will suffice. Various numerical methods can be used for obtaining such approximations. This chapter closely examines the key concepts of numerical differential equations and elucidates the finite difference method, numerical methods for ordinary differential equations and general linear methods.

Numerical Partial Differential Equations

Spectral Element Method

Spectral methods are essentially discretization methods for the approximate solution of partial-differential equations expressed in a weak form, based on high-order Lagrangian interpolants used in conjunction with particular quadrature rules. For spectral methods it is assumed that the solution can be expressed as a series of polynomial basis functions, which can approximate the solution well in some norm as the polynomial degree tends to infinity. These smooth basis-functions usually form an L2-complete basis. For the sake of computational efficiency this basis is typically chosen to be orthogonal in a weighted inner-product.

The spectral element method is a high-order finite element technique that combines the geometric flexibility of finite elements with the high accuracy of spectral methods.

It exhibits several favourable computational properties, such as the use of tensor products, naturally diagonal mass matrices, and adequacy to implementations in a parallel computer system. Due to these advantages, the spectral element method is a viable alternative to currently popular methods such as finite volumes and finite elements, if accurate solutions of regular problems are sought.

Main Features of the Spectral Element Method

- Lagrange interpolants shape functions

  The method is implemented by using tensor-product Lagrange interpolants within each element, where the nodes of these shape functions are placed at the zeros of Legendre polynomials (Gauss-Lobatto points) mapped from the reference domain [-1, 1] x [-1, 1] to each element. For smooth functions it can be shown that the resulting interpolants converge exponentially fast as the order of the interpolant is increased.

- Gauss-Lobatto quadrature

  Efficiency is achieved by using Gauss-Lobatto quadrature for evaluating elemental integrals: the quadrature points reside at the nodal points, which enables fast tensor-product techniques to be used for iterative matrix solution methods. Gauss-Lobatto quadrature results naturally in diagonal mass matrices.

One-dimensional Spectral Element Method

We use the 1-dimensional elliptic Helmholtz equation as a model equation.

$$\begin{cases} \alpha\, u - u'' = f, & x \in I = (0, L), \\ u(0) = u(L) = 0. \end{cases}$$

The corresponding weak form is,

$$\begin{cases} \text{Find } u \in H_0^1(I), \text{ such that} \\ \alpha\,(u, v) + (u', v') = (f, v), & \forall\, v \in H_0^1(I). \end{cases}$$

Spectral Element Method in Higher-dimensions

The spectral element method in 2-dimensional domains and 3-dimensional domains are similar. For simplicity, we only consider the 2-dimensional case here. The model equation we will use,

$$\begin{cases} \alpha\, u - \Delta u = f, & (x_1, x_2) \in \Omega \\ u|\partial\Omega = 0. \end{cases}$$

Here, we do not need Ω to be tensor product domain, since we can partition it into several small elements. These small elements can be quadrilateral elements, triangular elements or curved quadrangles or triangles. Here, we consider both quadrilateral and triangular elements. The weak form of above equation is,

$$\begin{cases} \text{Find } u \in H_0^1(\Omega), \text{ such that} \\ \alpha\,(u, v) + (\nabla u, \nabla v) = (f, v), & \forall\, v \in H_0^1(\Omega). \end{cases}$$

Finite Difference Method

The principle of finite difference methods is close to the numerical schemes used to solve ordinary differential equations. It consists in approximating the differential operator by replacing the derivatives in the equation using differential quotients. The domain is partitioned in space and in time and approximations of the solution are computed at the space or time points. The error between the numerical solution and the exact solution is determined by the error that is commited by going from a differential operator to a difference operator. This error is called the discretization error or truncation error. The term truncation error reflects the fact that a finite part of a Taylor series is used in the approximation.

For the sake of simplicity, we shall consider the one-dimensional case only. The main concept behind any finite difference scheme is related to the definition of the derivative of a smooth function u at a point $x \in \mathbb{R}$:

$$u'(x) = \lim_{h \to 0} \frac{u(x+h) - u(x)}{h},$$

and to the fact that when h tends to 0 (without vanishing), the quotient on the right-hand side provides a "good" approximation of the derivative. In other words, h should be sufficiently small to get a good approximation. It remains to indicate what exactly is a good approximation, in what sense. Actually, the approximation is good when the error commited in this approximation (i.e. when replacing the derivative by the differential quotient) tends towards zero when h tends to zero. If the function u is sufficiently smooth in the neighborhood of x, it is possible to quantify this error using a Taylor expansion.

Taylor Series

Suppose the function u is C^2 continuous in the neighborhood of x. For any h > 0 we have:

$$u(x+h) = u(x) + hu'(x) + \frac{h^2}{2} u''(x+h_1)$$

where h_1 is a number between 0 and h (i.e. x + h_1 is point of $[x, x+h]$). For the treatment of problems, it is convenient to retain only the first two terms of the previous expression:

$$u(x+h) = u(x) + hu'(x) + O(h^2)$$

where the term $O(h^2)$ indicates that the error of the approximation is proportional to h^2. From the equation ($u(x+h) = u(x) + hu'(x) + \frac{h^2}{2} u''(x+h_1)$), we deduce that there exists a constant $C > 0$, such that for $h > 0$ sufficiently small we have:

$$\left| \frac{u(x+h) - u(x)}{h} - u'(x) \right| \le C h, \quad C \sup_{y \in [x, x+h_0]} \frac{|u''(y)|}{2},$$

for $h \leq h_o$ ($h_o > 0$ given). The error committed by replacing the derivative $u'(x)$ by the differential quotient is of order h. The approximation of u' at point x is said to be consistent at the first order. This approximation is known as the forward difference approximant of u'. More generally, we define an approximation at order p of the derivative.

The approximation of the derivative u' at point x is of order p (p > 0) if there exists a constant C > 0, independent of h, such that the error between the derivative and its approximation is bounded by Ch^p (i.e. is exactly $O(h^p)$).

Likewise, we can define the first order backward difference approximation of u' at point x as:

$$u(x - h) = u(x) - hu'(x) + O(h^2).$$

Obviously, other approximations can be considered. In order to improve the accuracy of the approximation, we define a consistent approximation, called the central difference approximation, by taking the points $x - h$ and $x + h$ into account. Suppose that the function u is three times differentiable in the vicinity of x:

$$u(x + h) = u(x) + hu'(x) + \frac{h^2}{2}u''(x) + \frac{h^3}{6}u^{(3)}(\xi^+)$$

$$u(x + h) = u(x) + hu'(x) + \frac{h^2}{2}u''(x) + \frac{h^3}{6}u^{(3)}(\xi^-)$$

where $\xi^+ \in [x, x+h]$ and $\xi^- \in [x-h, x]$. By subtracting these two expressions we obtain, thanks to the intermediate value theorem:

$$\left| \frac{u(x + h) - u(x - h)}{2h} - u'(x) \right| \leq C h^2, \quad C = \sup_{y \in [x-h_0, x+h_0]} \frac{\left| u^{(3)(y)} \right|}{6}.$$

This defines a second order consistent approximation to u'

Remark: The order of the approximation is related to the regularity of the function u. If u is C^2 continuous, then the approximation is consistent at the order one only.

Approximation of the Second Derivative

Lemma: Suppose u is a C^4 continuous function on an interval $[x - h_o, x + h_o]$, $h_o > 0$. Then, there exists a constant C > 0 such that for every $h \in [0, h_o]$ we have:

$$\left| \frac{u(x + h) - 2u(x) + u(x - h)}{h^2} - u''(x) \right| \leq C h^2.$$

The differential quotient $\dfrac{u(x + h) - 2u(x) + u(x - h)}{h^2}$ is a consistent second-order approximation of the second derivative u'' of u at point x.

Proof. We use Taylor expansions up to the fourth order to achieve the result:

$$u(x+h) = u(x) + hu'(x) + \frac{h^2}{2}u''(x) + \frac{h^3}{6}u^{(3)}(x) + \frac{h^4}{24}u^{(4)}(\xi^+)$$

$$u(x-h) = u(x) - hu'(x) + \frac{h^2}{2}u''(x) - \frac{h^3}{6}u^{(3)}(x) + \frac{h^4}{24}u^{(4)}(\xi^-)$$

where $\xi^+ \in [x, x+h]$ and $\xi^- \in [x-h, x]$. Like previously, the intermediate value theorem allows us to write:

$$\frac{u(x+h) - 2u(x) + u(x-h)}{h^2} - u''(x) + \frac{h^2}{12}u^{(4)}(\xi),$$

where $\xi \in [x-h, x+h]$. Hence, we deduce the relation
$$\left| \frac{u(x+h) - 2u(x) + u(x-h)}{h^2} - u''(x) \right| \leq C h^2. \text{ with the constant}$$

$$C = \sup_{y \in [x-h_0, x+h_0]} \frac{\left| u^{(4)}(y) \right|}{12}.$$

Remark: Likewise, the error estimate depends on the regularity of the function u. If u is C^3 continuous, then the error is of order h only.

Example: The heat equation,

Consider the normalized heat equation in one dimension, with homogeneous Dirichlet boundary conditions

$$U_t = U_{xx}$$

$$U(0,t) = U(1,t) = 0 \quad \text{(boundary condition)}$$

$$U(x,0) = U_0(x) \quad \text{(initial condition)}$$

One way to numerically solve this equation is to approximate all the derivatives by finite differences. We partition the domain in space using a mesh $x_0, ..., x_J$ and in time using a mesh $t_0, ..., t_N$. We assume a uniform partition both in space and in time, so the difference between two consecutive space points will be h and between two consecutive time points will be k. The points

$$u(x_j, t_n) = u_j^n$$

will represent the numerical approximation of $u(x_j, t_n)$.

Explicit Method

Using a forward difference at time   and a second-order central difference for the space derivative at position x_j (FTCS) we get the recurrence equation:

$$\frac{u_j^{n+1} - u_j^n}{k} = \frac{u_{j+1}^n - 2u_j^n + u_{j-1}^n}{h^2}.$$

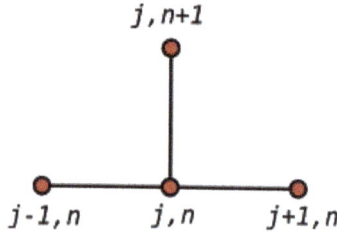

j, n+1

j-1, n   *j, n*   *j+1, n*

The stencil for the most common explicit method for the heat equation

This is an explicit method for solving the one-dimensional heat equation.

We can obtain u_j^{n+1} from the other values this way:

$$u_j^{n+1} = (1 - 2r)u_j^n + ru_{j-1}^n + ru_{j+1}^n$$

where $r = k/h^2$.

So, with this recurrence relation, and knowing the values at time n, one can obtain the corresponding values at time $n+1$. u_0^n and u_J^n must be replaced by the boundary conditions, in this example they are both 0.

This explicit method is known to be numerically stable and convergent whenever $r \leq 1/2$. The numerical errors are proportional to the time step and the square of the space step:

$$\Delta u = O(k) + O(h^2)$$

Implicit Method

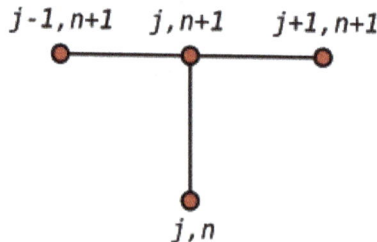

j-1, n+1   *j, n+1*   *j+1, n+1*

j, n

The implicit method stencil

If we use the backward difference at time t_{n+1} and a second-order central difference for the space derivative at position x_j (The Backward Time, Centered Space Method "BTCS") we get the recurrence equation:

$$\frac{u_j^{n+1} - u_j^n}{k} = \frac{u_{j+1}^{n+1} - 2u_j^{n+1} + u_{j-1}^{n+1}}{h^2}.$$

This is an implicit method for solving the one-dimensional heat equation.

We can obtain u_j^n from solving a system of linear equations:

$$(1+2r)u_j^{n+1} - ru_{j-1}^{n+1} - ru_{j+1}^{n+1} = u_j^n$$

The scheme is always numerically stable and convergent but usually more numerically intensive than the explicit method as it requires solving a system of numerical equations on each time step. The errors are linear over the time step and quadratic over the space step:

$$\Delta u = O(k) + O(h^2).$$

Crank–Nicolson method

Finally if we use the central difference at time $t_{n+1/2}$ and a second-order central difference for the space derivative at position x_j ("CTCS") we get the recurrence equation:

$$\frac{u_j^{n+1} - u_j^n}{k} = \frac{1}{2}\left(\frac{u_{j+1}^{n+1} - 2u_j^{n+1} + u_{j-1}^{n+1}}{h^2} + \frac{u_{j+1}^n - 2u_j^n + u_{j-1}^n}{h^2} \right).$$

This formula is known as the Crank–Nicolson method.

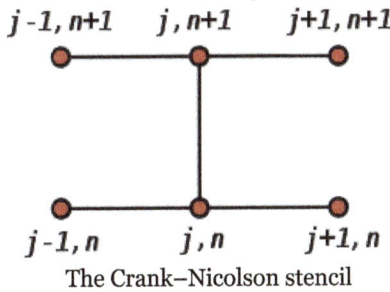

The Crank–Nicolson stencil

We can obtain u_j^{n+1} from solving a system of linear equations:

$$(2+2r)u_j^{n+1} - ru_{j-1}^{n+1} - ru_{j+1}^{n+1} = (2-2r)u_j^n + ru_{j-1}^n + ru_{j+1}^n$$

The scheme is always numerically stable and convergent but usually more numerically intensive as it requires solving a system of numerical equations on each time step. The errors are quadratic over both the time step and the space step:

$$\Delta u = O(k^2) + O(h^2).$$

Usually the Crank–Nicolson scheme is the most accurate scheme for small time steps. The explicit scheme is the least accurate and can be unstable, but is also the easiest to implement and the least numerically intensive. The implicit scheme works the best for large time steps.

Comparison

The figures below present the solutions given by the above methods to approximate the heat equation

$$U_t = \alpha U_{xx}, \quad \alpha = \frac{1}{\pi^2},$$

with the boundary condition

$$U(0,t) = U(1,t) = 0.$$

The exact solution is

$$U(x,t) = \frac{1}{\pi^2} e^{-t} \sin(\pi x).$$

Comparison of Finite Difference Methods

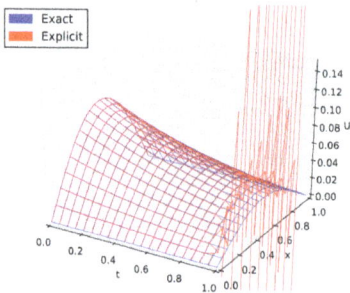

Explicit method (*not* stable)

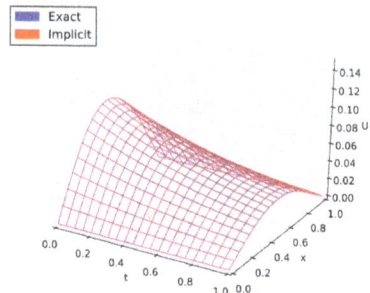

Implicit method (stable)

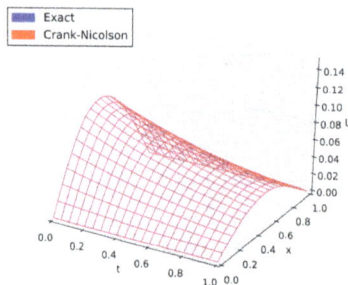

Crank-Nicolson method (stable)

Example: The Laplace Operator

The (continuous) Laplace operator in n-dimensions is given by $\Delta u(x) = \sum_{i=1}^{n} \partial_i^2 u(x)$. The discrete Laplace operator $\Delta_h u$ depends on the dimension n.

In 1D the Laplace operator is approximated as

$$\Delta u(x) = u''(x) \approx \frac{u(x-h) - 2u(x) + u(x+h)}{h^2} =: \Delta_h u(x).$$

This approximation is usually expressed via the following stencil

$$\frac{1}{h^2}\begin{bmatrix} 1 & -2 & 1 \end{bmatrix}.$$

The 2D case shows all the characteristics of the more general nD case. Each second partial derivative needs to be approximated similar to the 1D case

$$\Delta u(x,y) = u_{xx}(x,y) + u_{yy}(x,y)$$
$$\approx \frac{u(x-h,y) - 2u(x,y) + u(x+h,y)}{h^2} + \frac{u(x,y-h) - 2u(x,y) + u(x,y+h)}{h^2}$$
$$= \frac{u(x-h,y) + u(x+h,y) - 4u(x,y) + u(x,y-h) + u(x,y+h)}{h^2}$$
$$=: \Delta_h u(x,y),$$

which is usually given by the following stencil

$$\frac{1}{h^2}\begin{bmatrix} & 1 & \\ 1 & -4 & 1 \\ & 1 & \end{bmatrix}.$$

Consistency

Consistency of the above-mentioned approximation can be shown for highly regular functions, such as $u \in C^4(\Omega)$. The statement is

$$\Delta u - \Delta\ u = \ (h\ ).$$

To proof this one needs to substitute Taylor Series expansions up to order 3 into the discrete Laplace operator.

Properties

Subharmonic

Similar to continuous subharmonic functions one can define *subharmonic functions* for finite-difference approximations

$$-\Delta_h u_h \leq 0.$$

Mean value

One can define a general stencil of *positive type* via

$$\begin{bmatrix} & \alpha_N & \\ \alpha_W & -\alpha_C & \alpha_E \\ & \alpha_S & \end{bmatrix}, \quad \alpha_i > 0, \quad \alpha_C = \sum_{i \in \{N,E,S,W\}} \alpha_i.$$

If u_h is (discrete) subharmonic then the following *mean value property* holds

$$u_h(x_C) \leq \frac{\sum\limits_{i \in \{N,E,S,W\}} \alpha_i u_h(x_i)}{\sum\limits_{i \in \{N,E,S,W\}} \alpha_i},$$

where the approximation is evaluated on points of the grid, and the stencil is assumed to be of positive type.

A similar mean value property also holds for the continuous case.

Maximum Principle

For a (discrete) subharmonic function u_h the following holds

$$\max\nolimits_{\Omega_h} u_h \leq \max\nolimits_{\partial\Omega_h} u_h,$$

where $\Omega_h, \partial\Omega_h$ are discretizations of the continuous domain Ω, respectively the boundary $\partial\Omega$.

Numerical Methods for Ordinary Differential Equations

Euler Method

In mathematics and computational science, the Euler method (also called forward Euler method) is a first-order numerical procedurefor solving ordinary differential equations (ODEs) with a given initial value.

Let's start with a general first order IVP,

$$\frac{dy}{dt} = f(t,y) \quad y(t_o) = y_o$$

where $f(t,y)$ is a known function and the values in the initial condition are also known numbers. We know that if f and f_y re continuous functions then there is a unique solution to the IVP in some interval surrounding $t = t_0$. So, let's assume that everything is nice and continuous so that we know that a solution will in fact exist.

We want to approximate the solution to (1) near $t = t_0$. We'll start with the two pieces of information that we do know about the solution. First, we know the value of the solution at $t = t_0$. from the initial condition. Second, we also know the value of the derivative at $t = t_0$. We can get this by plugging the initial condition into $f(t,y)$ into the differential equation itself. So, the derivative at this point is.

$$\frac{dy}{dt}\bigg|_{t=t_0} = f(t_0, y_0)$$

Two pieces of information are enough for us to write down the equation of the tangent line to the solution at $t = t_0$. The tangent line is,

$$y = y_0 + f(t_0, y_0)(t - t_0)$$

Take a look at the figure below,

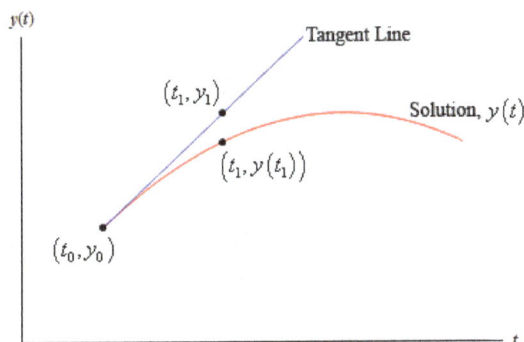

If t_1 is close enough to t_0 then the point y_1 on the tangent line should be fairly close to the actual value of the solution at t_1 or $y(t_1)$. Finding y_1 is easy enough. All we need to do is plug t_1 in the equation for the tangent line.

$$y_1 = y_0 + f(t_0, y_0)(t_1 - t_0)$$

Now, we would like to proceed in a similar manner, but we don't have the value of the solution at t_1 and so we won't know the slope of the tangent line to the solution at this point. This is a problem. We can partially solve it however, by recalling that y_1 is an approximation to the solution at t_1. If y_1 is a very good approximation to the actual value of the solution then we can use that to estimate the slope of the tangent line at t_1.

So, let's hope that y_1 is a good approximation to the solution and construct a line through the point (t_1, y_1) that has slope $f(t_1, y_1)$. This gives,

$$y = y_1 + f(t_1, y_1)(t - t_1)$$

Now, to get an approximation to the solution at $t = t_2$ we will hope that this new line will be fairly close to the actual solution at t_2 and use the value of the line at t_2 as an approximation to the actual solution. This gives.

$$y_2 = y_1 + f(t_1, y_1)(t_2 - t_1)$$

We can continue in this fashion. Use the previously computed approximation to get the next approximation. So,

$$y_3 = y_2 + f(t_2, y_2)(t_3 - t_2)$$
$$y_4 = y_3 + f(t_3, y_3)(t_4 - t_3)$$
$$\textit{etc.}$$

In general, if we have t_n and the approximation to the solution at this point, y_n and we want to find the approximation at t_{n+1} all we need to do is use the following.

$$y_{n+1} = y_n + f(t_n, y_n) \cdot (t_{n+1} - t_n)$$

If we define $f_n = f(t_n, y_n)$ we can simplify the formula to,

$$y_{n+1} = y_n + f_n \cdot (t_{n+1} - t_n)$$

Often, we will assume that the step sizes between the points $t_0, t_1, t_2, \ldots$ are of a uniform size of h. In other words, we will often assume that,

$$t_{n+1} - t_n = h$$

This doesn't have to be done and there are times when it's best that we not do this. However, if we do the formula for the next approximation becomes.

$$y_{n+1} = y_n + h f_n$$

Example

Given the initial value problem

$$y' = y, \quad y(0) = 1,$$

we would like to use the Euler method to approximate $y(4)$.

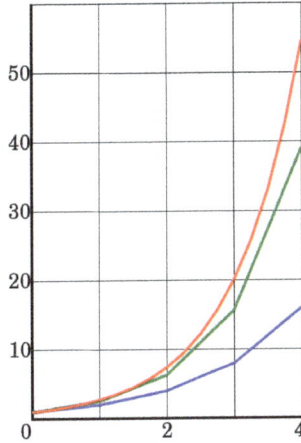

Numerical integration for the equation $y' = y, y(0) = 1$. Blue is the Euler method; green, the midpoint method; red, the exact solution, $y = e^t$. The step size is $h = 1.0$

Using Step Size Equal to 1 (h = 1)

The Euler method is

$$y_{n+1} = y_n + hf(t_n, y_n).$$

so first we must compute $f(t_o, y_o)$. In this simple differential equation, the function f is defined by $f(t, y) = y$. We have

$$f(t_o, y_o) = f(0, 1) = 1.$$

By doing the above step, we have found the slope of the line that is tangent to the solution curve at the point (0,1). Recall that the slope is defined as the change in y divided by the change in t, or $\Delta y / \Delta t$.

The next step is to multiply the above value by the step size h, which we take equal to one here:

$$h \cdot f(y_o) = 1 \cdot 1 = 1.$$

Since the step size is the change in t, when we multiply the step size and the slope of the tangent, we get a change in y value. This value is then added to the initial y value to obtain the next value to be used for computations.

$$y_o + hf(y_o) = y_1 = 1 + 1 \cdot 1 = 2.$$

The above steps should be repeated to find y_2, y_3 and y_4.

$$y_2 = y_1 + hf(y_1) = 2 + 1 \cdot 2 = 4,$$
$$y_3 = y_2 + hf(y_2) = 4 + 1 \cdot 4 = 8,$$
$$y_4 = y_3 + hf(y_3) = 8 + 1 \cdot 8 = 16.$$

Due to the repetitive nature of this algorithm, it can be helpful to organize computations in a chart form, as seen below, to avoid making errors.

n	y_n	t_n	$f(t_n, y_n)$	h	Δy	y_{n+1}
0	1	0	1	1	1	2
1	2	1	2	1	2	4
2	4	2	4	1	4	8
3	8	3	8	1	8	16

The conclusion of this computation is that $y_4 = 16$. The exact solution of the differential equation is $y(t) = e^t$, so $y(4) = e^4 \approx 54.598$. Thus, the approximation of the Euler method is not very good in this case. However, as the figure shows, its behaviour is qualitatively correct.

Using other Step Sizes

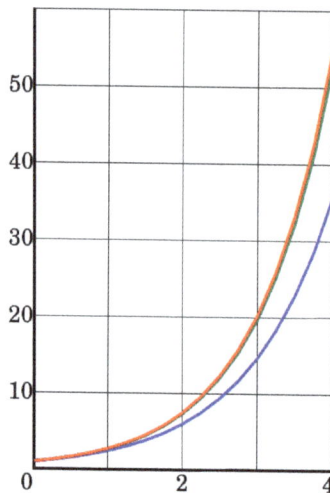

The same illustration for $h = 0.25$

The euler method is more accurate if the step size h is smaller. The table below shows the result with different step sizes. The top row corresponds to the example and the second row is illustrated in the figure.

Step size	Result of Euler's method	Error
1	16.00	38.60
0.25	35.53	19.07
0.1	41.15	13.45
0.05	49.56	5.04
0.025	51.98	2.62
0.0125	52.60	1.99

The error recorded in the last column of the table is the difference between the exact solution at $t = 4$ and the Euler approximation. In the bottom of the table, the step size is half the step size in the previous row, and the error is also approximately half the error in the previous row. This suggests that the error is roughly proportional to the step size, at least for fairly small values of the step size.

Other methods, such as the midpoint method also illustrated in the figures, behave more favourably: the global error of the midpoint method is roughly proportional to the *square* of the step size. For this reason, the Euler method is said to be a first-order method, while the midpoint method is second order.

We can extrapolate from the above table that the step size needed to get an answer that is correct to three decimal places is approximately 0.00001, meaning that we need 400,000 steps. This large number of steps entails a high computational cost. For this reason, people usually employ alternative, higher-order methods such as Runge–Kutta methods or linear multistep methods, especially if a high accuracy is desired.

Derivation

The Euler method can be derived in a number of ways. Firstly, there is the geometrical description above.

Another possibility is to consider the Taylor expansion of the function y around t_0:

$$y(t_0 + h) = y(t_0) + hy'(t_0) + \frac{1}{2}h^2 y''(t_0) + O(h^3).$$

The differential equation states that $y' = f(t, y)$. If this is substituted in the Taylor expansion and the quadratic and higher-order terms are ignored, the Euler method arises. The Taylor expansion is used below to analyze the error committed by the Euler method, and it can be extended to produce Runge–Kutta methods.

A closely related derivation is to substitute the forward finite difference formula for the derivative,

$$y'(t_0) \approx \frac{y(t_0 + h) - y(t_0)}{h}$$

in the differential equation $y' = f(t, y)$. Again, this yields the Euler method. A similar computation leads to the midpoint rule and the backward Euler method.

Finally, one can integrate the differential equation from t_0 to $t_0 + h$ and apply the fundamental theorem of calculus to get:

$$y(t_0 + h) - y(t_0) = \int_{t_0}^{t_0 + h} f(t, y(t))dt.$$

Now approximate the integral by the left-hand rectangle method (with only one rectangle):

$$\int_{t_0}^{t_0+h} f(t,y(t))dt \approx hf(t_0,y(t_0)).$$

Combining both equations, one finds again the Euler method. This line of thought can be continued to arrive at various linear multistep methods.

Local Truncation Error

The local truncation error of the Euler method is error made in a single step. It is the difference between the numerical solution after one step, y_1, and the exact solution at time $t_1 = t_0 + h$. The numerical solution is given by

$$y_1 = y_0 + hf(t_0,y_0).$$

For the exact solution, we use the Taylor expansion:

$$y(t_0 + h) = y(t_0) + hy'(t_0) + \frac{1}{2}h^2 y''(t_0) + O(h^3).$$

The local truncation error (LTE) introduced by the Euler method is given by the difference between these equations:

$$\text{LTE} = y(t_0 + h) - y_1 = \frac{1}{2}h^2 y''(t_0) + O(h^3).$$

This result is valid if y has a bounded third derivative.

This shows that for small h, the local truncation error is approximately proportional to h^2. This makes the Euler method less accurate (for small h) than other higher-order techniques such as Runge-Kutta methods and linear multistep methods, for which the local truncation error is proportional to a higher power of the step size.

A slightly different formulation for the local truncation error can be obtained by using the Lagrange form for the remainder term in Taylor's theorem. If y has a continuous second derivative, then there exists a $\xi \in [t_0, t_0 + h]$ such that

$$\text{LTE} = y(t_0 + h) - y_1 = \frac{1}{2}h^2 y''(\xi).$$

In the above expressions for the error, the second derivative of the unknown exact solution y can be replaced by an expression involving the right-hand side of the differential equation. Indeed, it follows from the equation $y' = f(t,y)$ that

$$y''(t_o) = \frac{\partial f}{\partial t}(t_o, y(t_o)) + \frac{\partial f}{\partial y}(t_o, y(t_o)) f(t_o, y(t_o)).$$

Global Truncation Error

The global truncation error is the error at a fixed time t, after however many steps the methods needs to take to reach that time from the initial time. The global truncation error is the cumulative effect of the local truncation errors committed in each step. The number of steps is easily determined to be $(t - t_o)/h$, which is proportional to $1/h$, and the error committed in each step is proportional to h^2. Thus, it is to be expected that the global truncation error will be proportional to h.

This intuitive reasoning can be made precise. If the solution y has a bounded second derivative and f is Lipschitz continuous in its second argument, then the global truncation error (GTE) is bounded by,

$$|\text{GTE}| \leq \frac{hM}{2L}(e^{L(t-t_o)} - 1)$$

where M is an upper bound on the second derivative of y on the given interval and L is the Lipschitz constant of

The precise form of this bound is of little practical importance, as in most cases the bound vastly overestimates the actual error committed by the Euler method. What is important is that it shows that the global truncation error is (approximately) proportional to h. For this reason, the Euler method is said to be first order.

Numerical Stability

The Euler method can also be numerically unstable, especially for stiff equations, meaning that the numerical solution grows very large for equations where the exact solution does not. This can be illustrated using the linear equation

$$y' = -2.3y, \qquad y(0) = 1.$$

The exact solution is $y(t) = e^{-2.3t}$, which decays to zero as $t \to \infty$. However, if the Euler method is applied to this equation with step size $h = 1,$, then the numerical solution is qualitatively wrong: it oscillates and grows. This is what it means to be unstable. If a smaller step size is used, for instance $h = 0.7,$, then the numerical solution does decay to zero.

If the Euler method is applied to the linear equation $y' = ky$, then the numerical solution is unstable if the product hk is outside the region

$$\{z \in C \,||\, z+1| \leq 1\},$$

Illustrated on the right. This region is called the (linear) instability region. In the example, k equals -2.3, so if $h = 1$ then $hk = -2.3$ which is outside the stability region, and thus the numerical solution is unstable.

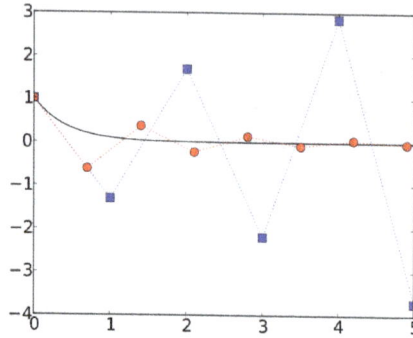

Solution of $y' = -2.3y$ computed with the Euler method with step size $h = 1$ (blue squares) and $h = 0.7$ (red circles). The black curve shows the exact solution

This limitation —along with its slow convergence of error with h— means that the Euler method is not often used, except as a simple example of numerical integration.

Rounding Errors

The discussion up to now has ignored the consequences of rounding error. In step n of the Euler method, the rounding error is roughly of the magnitude εy_n where ε is the machine epsilon. Assuming that the rounding errors are all of approximately the same size, the combined rounding error in N steps is roughly $N\varepsilon y_0$ if all errors points in the same direction. Since the number of steps is inversely proportional to the step size h, the total rounding error is proportional to ε / h. In reality, however, it is extremely unlikely that all rounding errors point in the same direction. If instead it is assumed that the rounding errors are independent rounding variables, then the total rounding error is proportional to $\varepsilon / \sqrt{h}$.

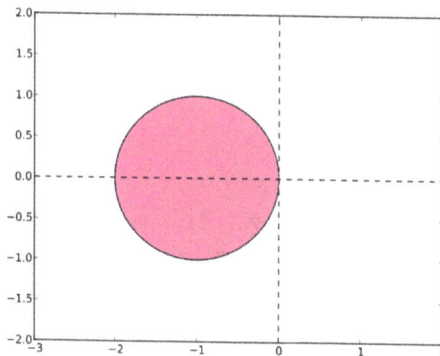

The pink disk shows the stability region for the Euler method

Thus, for extremely small values of the step size, the truncation error will be small but the effect of rounding error may be big. Most of the effect of rounding error can be easily avoided if compensated summation is used in the formula for the Euler method.

Modifications and Extensions

A simple modification of the Euler method which eliminates the stability problems is the backward Euler method:

$$y_{n+1} = y_n + hf(t_{n+1}, y_{n+1}).$$

This differs from the (standard, or forward) Euler method in that the function f is evaluated at the end point of the step, instead of the starting point. The backward Euler method is an implicit method, meaning that the formula for the backward Euler method has y_{n+1} on both sides, so when applying the backward Euler method we have to solve an equation. This makes the implementation more costly.

Other modifications of the Euler method that help with stability yield the exponential Euler method or the semi-implicit Euler method.

More complicated methods can achieve a higher order (and more accuracy). One possibility is to use more function evaluations.

$$y_{n+1} = y_n + hf\left(t_n + \tfrac{1}{2}h, y_n + \tfrac{1}{2}hf(t_n, y_n)\right).$$

This leads to the family of Runge–Kutta methods.

The other possibility is to use more past values, as illustrated by the two-step Adams–Bashforth method:

$$y_{n+1} = y_n + \tfrac{3}{2}hf(t_n, y_n) - \tfrac{1}{2}hf(t_{n-1}, y_{n-1}).$$

This leads to the family of linear multistep methods. There are other modifications which uses techniques from compressive sensing to minimize memory usage.

Heun Method

In mathematics and computational science, Heun's method may refer to the improved or modified Euler's method (that is, the explicit trapezoidal rule), or a similar two-stage Runge–Kutta method. It is named after Karl Heun and is a numerical procedure for solving ordinary differential equations (ODEs) with a given initial value. Both variants can be seen as extensions of the Euler method into two-stage second-order Runge–Kutta methods.

The procedure for calculating the numerical solution to the initial value problem via the improved Euler's method is:

$$y'(t) = f(t, y(t)), \qquad y(t_0) = y_0,$$

by way of Heun's method, is to first calculate the intermediate value $\tilde{y}_{i+1}$ and then the final approximation y_{i+1} at the next integration point.

$$\tilde{y}_{i+1} = y_i + hf(t_i, y_i)$$

$$y_{i+1} = y_i + \frac{h}{2}[f(t_i, y_i) + f(t_{i+1}, \tilde{y}_{i+1})],$$

where h is the step size and $t_{i+1} = t_i + h$.

Euler's method is used as the foundation for Heun's method. Euler's method uses the line tangent to the function at the beginning of the interval as an estimate of the slope of the function over the interval, assuming that if the step size is small, the error will be small. However, even when extremely small step sizes are used, over a large number of steps the error starts to accumulate and the estimate diverges from the actual functional value.

Where the solution curve is concave up, its tangent line will underestimate the vertical coordinate of the next point and vice versa for a concave down solution. The ideal prediction line would hit the curve at its next predicted point. In reality, there is no way to know whether the solution is concave-up or concave-down, and hence if the next predicted point will overestimate or underestimate its vertical value. The concavity of the curve cannot be guaranteed to remain consistent either and the prediction may overestimate and underestimate at different points in the domain of the solution. Heun's Method addresses this problem by considering the interval spanned by the tangent line segment as a whole. Taking a concave-up example, the left tangent prediction line underestimates the slope of the curve for the entire width of the interval from the current point to the next predicted point. If the tangent line at the right end point is considered (which can be estimated using Euler's Method), it has the opposite problem The points along the tangent line of the left end point have vertical coordinates which all underestimate those that lie on the solution curve, including the right end point of the interval under consideration. The solution is to make the slope greater by some amount. Heun's Method considers the tangent lines to the solution curve at *both* ends of the interval, one which *overestimates*, and one which *underestimates* the ideal vertical coordinates. A prediction line must be constructed based on the right end point tangent's slope alone, approximated using Euler's Method. If this slope is passed through the left end point of the interval, the result is evidently too steep to be used as an ideal prediction line and overestimates the ideal point. Therefore, the ideal point lies approximately halfway between the erroneous overestimation and underestimation, the average of the two slopes.

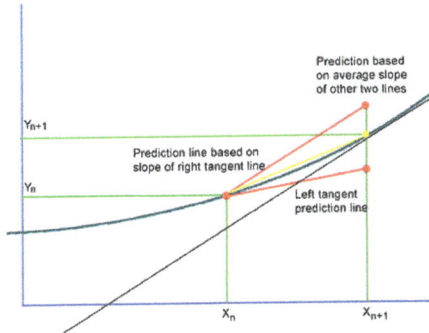

A diagram depicting the use of Heun's method to find a less erroneous prediction
when compared to the lower order Euler's Method

Euler's Method is used to roughly estimate the coordinates of the next point in the solution, and with this knowledge, the original estimate is re-predicted or *corrected*. Assuming that the quantity $f(x,y)$ on the right hand side of the equation can be thought of as the slope of the solution sought at any point (x,y),, this can be combined with the Euler estimate of the next point to give the slope of the tangent line at the right endpoint. Next the average of both slopes is used to find the corrected coordinates of the right end interval.

Derivation

$$\text{Slope}_{\text{left}} = f(z_i, y_i)$$

$$\text{Slope}_{\text{right}} = f(z_i + h, y_i + hf(z_i, y_i))$$

$$\text{Slope}_{\text{ideal}} = (1/2)(\text{Slope}_{\text{left}} + \text{Slope}_{\text{right}})$$

Using the principle that the slope of a line equates to the rise/run, the coordinates at the end of the interval can be found using the following formula:

$$\text{Slope}_{\text{ideal}} = \frac{\Delta y}{h}$$

$$\Delta y = h(\text{Slope}_{\text{ideal}})$$

$$x_{i+1} = x_i + h, \quad y_{i+1} = y_i + \Delta y$$

$$y_{i+1} = y_i + h\text{Slope}_{\text{ideal}}$$

$$y_{i+1} = y_i + \frac{1}{2}h(\text{Slope}_{\text{left}} + \text{Slope}_{\text{right}})$$

$$y_{i+1} = y_i + \frac{h}{2}(f(x_i, y_i) + f(x_i + h, y_i + hf(x_i, y_i)))$$

The accuracy of the Euler method improves only linearly with the step size is decreased, whereas the Heun Method improves accuracy quadratically . The scheme can be compared with the implicit trapezoidal method, but with $f(t_{i+1}, y_{i+1})$ replaced by $f(t_{i+1}, \tilde{y}_{i+1})$ in order to make it explicit. $\tilde{y}_{i+1}$ is the result of one step of Euler's method on the same initial value problem. So, Heun's method is a predictor-corrector method with forward Euler's method as predictor and trapezoidal method as corrector.

Runge–Kutta method

The improved Euler's method is a two-stage Runge–Kutta method, and can be written using the Butcher tableau (after John C. Butcher):

$$
\begin{array}{c|cc}
0 & & \\
1 & 1 & \\
\hline
& 1/2 & 1/2
\end{array}
$$

The other method referred to as Heun's method (also known as Ralston's method) has the Butcher table:

$$
\begin{array}{c|cc}
0 & & \\
2/3 & 2/3 & \\
\hline
& 1/4 & 3/4
\end{array}
$$

This method minimizes the truncation error.

The solution of differential equation with desired accuracy can be achieved using classical Taylor series method at a specified point. This means for given h, one can go on adding more and more terms of the series till the desired accuracy is achieved. This requires the expressions for several higher order derivatives and its evaluation. It poses practical difficulties in the application of Taylor series method:

- Higher order derivatives may not be easily obtained.

- Even if the expressions for derivatives are obtained, lot of computational effort may still be required in their numerical evaluation.

It is possible to develop one step algorithms which require evaluation of first derivative as in Euler method but yields accuracy of higher order as in Taylor series. These methods require functional evaluations of $f(t,y(t))$ at more than one point.on the interval $[t_k, t_{k+1}]$. The Category of methods are known as Runge-Kutta methods of order 2, 3 and more depending upon the order of accuracy. A general Runge Kutta algorithm is given as,

$$y_{k+1} = y_k + h\phi(t_k, y_k, h)$$

The function phi is termed as increment function. The mth order Runge-Kutta method gives accuracy of order h^m. The function ϕ is chosen in such a way so that when expanded the right hand side of ($y_{k+1} = y_k + h\phi(t_k, y_k, h)$) matches with the Taylor series upto desired order. This means that for a second order Runge-Kutta mehod the right side of ($y_{k+1} = y_k + h\phi(t_k, y_k, h)$) matches up to second order terms of Taylor series.

Second Order Runge Kutta Methods

The Second order Runge Kutta methods are known as RK2 methods. For the derivation of second order Runge Kutta methods, it is assumed that phi is the weighted average of two functional evaluations at suitable points in the interval $[t_k, t_{k+1}]$:

$$\phi(t_k, y_k, h) = w_1 K_1 + w_2 K_2$$
$$K_1 = f(t_k, y_k)$$
$$K_2 = f(t_k + ph, y_k + qhK_1) : 0 \le p, q \le 1$$

Here, four constants w_1, w_2, p and q are introduced. These are to be chosen in such a way that the expansion matches with the Taylor series up to second order terms.

For this,

$$K_2 = f(t_k + ph, y_k + qhK_1)$$
$$= f(t_k, y_k) + phf_t(t_k, y_k) + qhK_1 f_y(t_k, y_k) + O(h^2)$$
$$f(t_k, y_k) + phf_t(t_k, y_k) + qhf(t_k, y_k)(f_y(t_k, y_k) + O(h^2)$$

Substitution in ($y_{k+1} = y_k + h\phi(t_k, y_k, h)$)) yields,

$$y_{k+1} = y_k + h[w_1 f(t_k, y_k) + w_2 \{f(t_k, y_k) + phf_t(t_k, y_k) + qhf_t(t_k, y_k)(f_y(t_k, y_k) + O(h^2)\}]$$

or

$$y_{k+1} = y_k + h[w_1 f(t_k, y_k) + w_2 f(t_k, y_k)] + h^2[pf_t(t_k, y_k) + qf(t_k, y_k)(f_y(t_k, y_k) + O(h^3))$$

Using chain rule for the derivative $f'(t_k, y(t_k))$ gives,

$$f'(t_k, y(t_k)) = f_t(t_k, y(t_k)) + f(t_k, y(t_k))(f_y(t_k, y(t_k))$$

Substituting in Taylor series gives,

$$y(t_k + h) = y(t_k) + hf(t_k + y(t_k)) + \frac{h^2}{2}[f_t(t_k, y(t_k)) + f(t_k, y(t_k))(f_y(t_y, y(t_k))] + O(h^3)$$

Assuming $y(t_k) = y_k$ and comparing

$$y_{k+1} = y_k + h[w_1 f(t_k, y_k) + w_2 f(t_k, y_k)] + h^2[pf_t(t_k, y_k) + qf(t_k, y_k)(f_y(t_k, y_k) + O(h^3) \text{ and}$$

$$y(t_k + h) = y(t_k) + hf(t_k + y(t_k)) + \frac{h^2}{2}[f_t(t_k, y(t_k)) + f(t_k, y(t_k))(f_y(t_y, y(t_k))] + O(h^3)$$

yields,

$$w_1 + w_2 = 1, \ w_1 p = 1/2 \ \text{and} \ w_2 q = 1/2$$

Observe that four unknowns are to be evaluated from three equations. Accordingly many solutions are possible for above equation. Let us chose arbitrary value to constant q as q=1, then,

$$w_1 + w_2 = 1/2, \ p = 1 \ \text{and} \ q = 1$$

Accordingly, the second order Runge-Kutta can be written as,

$$y_{kp} = y_k + hf(t_k, y_k)$$

$$y_{k+1} = y_k + \frac{h}{2}[f(t_k, y_k) + f(t_k + h, y_{kp})]$$

This is the same as modified Euler method. It may be noted that the method reduces to a quadrature formula [Trapezoidal rule] when f(t, y) is independent of y:

$$y_{k+1} = y_k + \frac{h}{2}[f(t_k) + f(t_k + h)]$$

For convenience q is chosen between 0 and 1such that one of the weights w in the method is zero. For example choose q=1/2 makes w1=0 and (1.12) yields:

$$w_1 = 0, w_2 = 1, \ p = q = 1/2$$

$$\bar{y}_k = y_k + \frac{h}{2} f(t_k, y_k)$$

$$y_{k+1} = y_k + hf(t_k + \frac{h}{2}, \bar{y}_k)]$$

Choosing arbitrary constant q so as to minimize the sum of absolute values of coefficients in the truncation error term T_{j+1} gives optimal RK method. The minimum error occurs for q=2/3. Accordingly optimal method is obtained for,

$$w_1 = 1/4, w_2 = 3/4, \ p = q = 2/3$$

This gives another second order Runge-Kutta method known as optimal RK2 method:

$$\tilde{y}_k = y_k + \frac{2h}{3} f(t_k, y_k)$$

$$y_{k+1} = y_k + \frac{h}{4}f(t_k, y_k) + \frac{3h}{4}f(t_k + \frac{2h}{3}, \tilde{y}_k)$$

Example: Solve IVP in 1<t<2 with h=0.1using Optimal Runge Kutta Method above equation

$$y' = y/t - (y/t)^2; y(1) = 1$$

Solution: The solution is given in table below,

Table: Solution of Example above with h=0.1.

t	yk	f(t,y)	yk0	t+2h/3	f(t+2h/3,yk0)	yk+1
1	1	0	1	1.066667	0.05859375	1.004395
1.1	1.004395	0.07936	1.020267	1.166667	0.119744262	1.015359
1.2	1.015359	0.130192	1.041398	1.266667	0.159037749	1.030542
1.3	1.030542	0.164312	1.063404	1.366667	0.185456001	1.048559
1.4	1.048559	0.188014	1.086162	1.466667	0.203806563	1.068545
1.5	1.068545	0.204902	1.109525	1.566667	0.216857838	1.089932
1.6	1.089932	0.217164	1.133364	1.666667	0.226296619	1.112333
1.7	1.112333	0.226187	1.157571	1.766667	0.233198012	1.135478
1.8	1.135478	0.232886	1.182055	1.866667	0.238272937	1.15917
1.9	1.15917	0.23788	1.206746	1.966667	0.242006106	1.183268
2	1.183268	0.241603	1.231588	2.066667	0.244736661	1.207663

Fourth Order Runge-Kutta Methods (RK4)

All the fourth order Runge Kutta Methods are of the following general form:

$$y_{k+1} = y_k + h\phi(t_k, y_k, h)$$
$$\phi = w_1 K_1 + w_2 K_2 + w_3 K_3 + w_4 K_4$$
$$K_1 = f(t_k, y_k)$$
$$K_2 = f(t_k + p_1 h, y_k + a_{21} K_1)$$

$$K_3 = f(t_k + p_2 h, y_k + a_{31} K_1 + a_{22} K_2)$$
$$K_4 = f(t_k + p_3 h, y_k + a_{41} K_1 + a_{42} K_2 + a_{43} K_3)$$

The thirteen unknowns in the method have to be obtained. The Taylor series expansion of the solution and Ki, i=1,2,3,4 are obtained and substituted in the first equation of ($K_2 = f(t_k + p_1 h, y_k + a_{21} K_1)$)). For Fourth order Runge Kutta Method, comparing terms up to h[4] on the two sides gives the following 11 equations:

$$P_1 = a_{21}; P_2 = a_{31} + a_{32}; P_3 = a_{41} + a_{42} + a_{43}$$
$$w_1 + w_2 + w_3 + w_4 = 1$$
$$P_1 w_2 + P_2 w_3 + P_3 w_4 = 1/2$$
$$P_1^2 w_2 + P_2^2 w_3 + P_3^2 w_4 = 1/3$$
$$P_1 a_{32} w_2 + (P_2 a_{42} + P_2 a_{43}) w_4 = 1/6$$
$$P_1^3 w_2 + P_2^3 w_3 + P_3^3 w_4 = 1/4$$
$$a_{32} P_1^2 w_3 + (P_1^2 a_{42} + P_2 a_{43}) w_4 = 1/12$$
$$a_{32} P_1 P_2 w_3 + P_3 (P_1 a_{42} + P_2^2 a_{43}) w_4 = 1/8$$
$$P_1 a_{32} a_{43} w_4 = 1/24$$

Assuming two additional constraints $P_1 = P_2$ and $w_2 = w_3$ gives,

$$P_1 = P_2 = \frac{1}{2}, P_3 = 1, w_2 = w_3 = \frac{1}{3}$$

$$w_1 = w_4 = \frac{1}{6}, a_{21} = a_{32} = \frac{1}{2}$$

$$a_{31} = a_{41} = a_{42} = 0, a_{43} = 1$$

Accordingly the classical fourth order Runge-Kutta method is obtained as,

$$y_{k+1} = y_k + \frac{h}{6}[K_1 + 2K_2 + 2K_3 + K_4]$$

$$K_1 = f(t_k, y_k)$$

$$K_2 = f(t_k + \frac{h}{2}, y_k + \frac{h}{2}K_1)$$

$$K_3 = f(t_k + \frac{h}{2}, y_k + \frac{h}{2}K_2)$$

$$K_4 = f(t_k + h, y_k + hK_3)$$

It may be observed that RK4 uses four functional evaluations in the interval $[t_0, t_1]$.

These points are shown as p0, p1, p2 and p3 in the following figure.

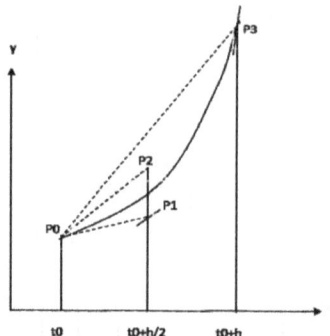

Figure: RK4 Method

Example: Find the solution of IVP using classical fourth order Runge-Kutta method with h=1,

$$\frac{dy}{dt} = \frac{1}{2} + \frac{y}{t}; y(1) = 0$$

Solution: The solution of IVP by RK4 classical method is shown in the following table:

Table: Solution by Classical RK4 for example above.

h	k	tk	yk	k1	k2	k3	k4	yk+1	exact sol	Abs error
1	0	1	0	0.5	0.66667	0.72222	0.861111	0.689815	0.693147	0.00333
1	1	2	0.68981	0.84491	0.94491	0.96491	1.051574	1.6425	1.647918	0.00542
1	2	3	1.6425	1.0475	1.11893	1.12913	1.192908	2.765255	2.772589	0.00733
1	3	4	2.76526	1.19131	1.24687	1.25304	1.303659	4.014388	4.023595	0.00921
1	4	5	4.01439	1.30288	1.34833	1.35246	1.394475	5.364212	5.375278	0.01107
1	5	6	5.36421	1.39404	1.4325	1.43546	1.471381	6.797766	6.810686	0.01292
1	6	7	6.79777	1.47111	1.50444	1.50666	1.538054	8.302995	8.317766	0.01477
1	7	8	8.303	1.53787	1.56729	1.56902	1.59689	9.87089	9.887511	0.01662
1	8	9	9.87089	1.59677	1.62308	1.62447	1.649536	11.49446	11.51293	0.01847
1	9	10	11.4945	1.64945	1.67326	1.67439	1.697168	13.16811	13.18842	0.02032
1	10	11	13.1681	1.6971	1.71884	1.71978	1.740658	14.88727	14.90944	0.02217

The other RK4 formulae are obtained as,

$$y_{k+1} = y_k + \frac{h}{8}[K_1 + 3K_2 + 3K_3 + K_4]$$

$$K_1 = f(t_k, y_k)$$

$$K_2 = f(t_k + \frac{h}{3}, y_k + \frac{h}{3}K_1)$$

$$K_3 = f(t_k + \frac{2h}{3}, y_k + \frac{h}{3}K_1 + hK_2)$$

$$K_4 = f(t_k + h, y_k + hK_1 - hK_2 + hK_3)$$

and

$$y_{k+1} = y_k + \frac{h}{6}[K_1 + 2(1 - \frac{1}{\sqrt{2}})K_2 + 2(1 + \frac{1}{\sqrt{2}})K_3 + K_4]$$

$$K_1 = f(t_k, y_k)$$

$$K_2 = f(t_k + \frac{h}{2}, y_k + \frac{h}{2}K_1)$$

$$K_3 = f(t_k + \frac{h}{2}, y_k + (-\frac{1}{\sqrt{2}} + \frac{1}{\sqrt{2}})hK_1 + (1 - \frac{1}{\sqrt{2}})hK_2)$$

$$K_4 = f(t_k + h, y_k - \frac{1}{\sqrt{2}}hK_2 + (1 + \frac{1}{\sqrt{2}})hK_3)$$

Example: Find the solution of IVP using classical fourth order Runge-Kutta method given in ($K_2 = f(t_k + \frac{h}{3}, y_k + \frac{h}{3}K_1)$) with h=1

$$\frac{dy}{dt} = \frac{1}{2} + \frac{y}{t}; y(1) = 0$$

Solution: In the following table the computation are shown to solve the IVP using ($K_2 = f(t_k + \frac{h}{3}, y_k + \frac{h}{3}K_1)$). Although both the methods are of same order but ($K_2 = f(t_k + \frac{h}{3}, y_k + \frac{h}{3}K_1)$) gives more accurate result as compared to classical method ($K_2 = f(t_k + p_1 h, y_k + a_{21}K_1)$)

Table: Solution by classical RK4 for example above.

h	k	tk	yk	k1	k2	k3	k4	yk+1	exact sol	Abs error
1	0	1	0	0.5	0.625	0.775	0.825	0.690625	0.693147	0.00252
1	1	2	0.69063	0.84531	0.91674	0.9971	1.038765	1.643824	1.647918	0.00409
1	2	3	1.64382	1.04794	1.09794	1.15249	1.186578	2.76705	2.772589	0.00554
1	3	4	2.76705	1.19176	1.23022	1.27143	1.300004	4.016642	4.023595	0.00695
1	4	5	4.01664	1.30333	1.33458	1.36767	1.392176	5.366922	5.375278	0.00836
1	5	6	5.36692	1.39449	1.4208	1.44843	1.469863	6.80093	6.810686	0.00976
1	6	7	6.80093	1.47156	1.49429	1.518	1.537026	8.306613	8.317766	0.01115
1	7	8	8.30661	1.53833	1.55833	1.5791	1.59619	9.874961	9.887511	0.01255
1	8	9	9.87496	1.59722	1.61508	1.63355	1.649065	11.49898	11.51293	0.01395
1	9	10	11.499	1.6499	1.66603	1.68266	1.696865	13.17308	13.18842	0.01534
1	10	11	13.1731	1.69755	1.71226	1.72738	1.74048	14.8927	14.90944	0.01674

Let $T_{k+1}(h)$ be the local truncation error at the (k+1)th step of the one step method with step size h, assuming that no error was made in the previous step. It is obtained as

$$T_{k+1}(h) = y_{k+1} - y_k - h\phi(t_k, y_k, h)$$

The method is said to be consistent if,

$$\lim_{h \to 0} \left(\frac{T_{k+1}(h)}{h} \right) = 0$$

It is now easy to verify that the Euler, Modified Euler and Runge-Kutta methods are consistent.

It is now easy to verify that the Eulers, Modified Eulers and Runge Kutta methods are consistent.

A one step method is convergent when the difference between the exact solution and the solution of difference equation at k^{th} step satisfies the condition,

$$\lim_{h \to 0} \left(\lim_{1 \le k \le N} |y(t_k) - y_k| \right) = 0$$

Using the bound for $t_k = y(t_k) - y_k$ proves the convergence of Euler method.

Higher Order Runge Kutta Methods

The local error of RK4 is $O(h^5)$ while the global error is $O(h^4)$. If the solution of IVP is to be obtained for t in the interval (0,1] with h=0.1 using RK4, then the solution is to be computed at 10 grid points t=0.1, 0.2, ... ,1.0. At each grid point four functional evaluations are required. This way, the solution at t=1 requires 40 functional evaluations and the accuracy will be of order 10^{-4}. If Euler method is to be used then h=0.0001 will yield the desired accuracy of 10^{-4}. Accordingly, the solution is to be computed at 10000 grid points to reach to t=1.This means Euler method requires10,000 functional evaluations (one for each grid point) to compute approximate solution at t=1. Evidently, RK4 is very efficient and saves lots of computational effort.

We can further improve the efficiency by employing still higher order Runge-Kutta methods. The higher order methods, say of order 5 and 6 are developed on the same lines. These are more efficient as the higher accuracy is achieved with less computational effort as compared to lower order methods.

Runge-Kutta method (say of order 4) is applied for obtaining approximate solution y_1 at $t=t_1=t+h$ for IVP with chosen value of h. The value of h is then halved and the solution is again obtained at t_1. Now 7 functional evaluations are needed to compute y_1. If the difference between two solutions is not substantial then the approximation is accepted. Otherwise, the iteration (halving h) is repeated again till the desired accuracy is achieved. Each halving h will require 4+7 functional evaluations. This way higher accuracy can be achieved with more computational effort.

In another approach, higher performance with less computational effort is achieved when the Runge-Kutta methods of different orders are used to move from one grid point to the next point. One such method known as Runge-Kutta Fehlberg method is based on the formulae (Error = $\hat{y}_{k+1} - \hat{y}_{k+1} = [\frac{1}{360} K_1 - \frac{128}{4275} K_3 - \frac{2197}{75240} K_4 + \frac{1}{50} K_5 + \frac{2}{55} K_6]h$) given below. In this, two estimates are obtained for y_{k+1} using RK method of global error $O(h^4)$ and $O(h^5)$ with six functional evaluations.

$$K_1 = f(t_k, y_k)$$

$$K_2 = f(t_k + \frac{h}{4}, y_k + \frac{h}{4}K_1)$$

$$K_3 = f(t_k + \frac{3h}{8}, y_k + \frac{3h}{32}K_1 + \frac{9}{32}K_2)$$

$$K_4 = f(t_k + \frac{12h}{13}, y_k + \frac{1932h}{2197}K_1 - \frac{7200h}{2197}K_2 + \frac{7296h}{2197}K_3)$$

$$K_5 = f(t_k + h, y_k + \frac{439h}{216}K_1 - 8hK_2 + \frac{3680h}{513}K_3 - \frac{845h}{4104}K_4)$$

$$K_6 = f(t_k + \frac{h}{2}, y_k - \frac{8h}{27}K_1 + 2hK_2 - \frac{3544h}{2565}K_3 + \frac{1859h}{4104}K_4 - \frac{11h}{40}K_5)$$

$$\widehat{y}_{k+1} = y_k + \frac{h}{8}[\frac{25}{216}K_1 + \frac{1408}{2565}K_3 + \frac{2197}{4104}K_4 - \frac{1}{5}K_5]$$

$$\widehat{y}_{k+1} = y_k + \frac{h}{8}[\frac{16}{135}K_1 + 12825K_3 + \frac{28561}{56430}K_4 - \frac{9}{50}K_5 + \frac{2}{55}K_6]$$

$$\text{Error} = \widehat{y}_{k+1} - \widehat{y}_{k+1} = [\frac{1}{360}K_1 - \frac{128}{4275}K_3 - \frac{2197}{75240}K_4 + \frac{1}{50}K_5 + \frac{2}{55}K_6]h$$

Since the lower order method is of order four, the step size adjustment factor s can be computed as

$$s \leq 0.84\left(\frac{\varepsilon}{|T_{i+1}|}\right)^{1/4}$$

Here, ε is the accuracy requirement and T is the truncation error $T_{i+1} = |y_{i+1} - \widehat{y}_{i+1}|$.

If the desired accuracy is not achieved the solution is iterated taking new value of h. Depending upon error requirement the step size h can be increased or decreased. The solution y_{k+1} of desired accuracy is obtained at tk+1=tk+sh. The method is known as RKF45. To implement the method, the user specifies the allowable smallest step size hmin , largest step size hmax and the maximum allowable local truncation error ε. The following algorithm is used to solve IVP using RKF45 formulae with self adjusting variable step sizes:

Algorithm RKF45

- [Step 1] set k=0, t=a=t$_0$, y=yk, h=h$_{max}$, flag=1

- [Step 2] while (flag==1) repeat steps 3-7

- [Step 3] compute $\widehat{y}_{k+1}, \widehat{y}_{k+1}$ and $|\widehat{y}_{k+1} - \widetilde{y}_{k+1}|$

- [Step 4] compute s
- [Step 5] if s >h$_{max}$ then h=h$_{max}$ else if (s<h$_{min}$) exit else h=s
- [Step 6] if $(R \leq \varepsilon)$ flag=0
- [Step 7] go to step 2
- [Step 8] t=t+h, y=yk+1, k++; flag=1
- [Step 9] if(t<=b) goto step 2
- [Step 10] stop

Example: Solve IVP using $y' = 2ty^2$, $y(0)=1$.

Solution: the matlab code for solving the system is given as

```
function dydt= rkf4(t,y)
dydt=y.^2.2*t;
[t,y]=ode45(@rkf4,[0 1], 1); %[0 1]-time span, 1-initial condition
```

Table: Solution of example 1.8

t	y
0.1	1.005
0.2	1.02
0.3	1.047
0.4	1.088
0.5	1.145
0.6	1.225
0.7	1.337
0.8	1.497
0.9	1.741
1.0	2.146

Example: find the solution of IVP using higher order order Runge-Kutta method RKF45 given in (Error $= \hat{y}_{k+1} - \hat{y}_{k+1} = [\dfrac{1}{360}K_1 - \dfrac{128}{4275}K_3 - \dfrac{2197}{75240}K_4 + \dfrac{1}{50}K_5 + \dfrac{2}{55}K_6]h$) in the interval $(0,2)$ taking h max=0.25, hmin=0.01 and accuracy as 0.000001

$$\frac{dy}{dt} = y - t^2 + 1; y(0) = 0.5$$

Solution: The detailed solution of the problem is worked out in the excel sheet rkf45.xls

Table: Details of Solution of example 1.9

h	t	y	k1	k2 k3	k4	k5	k6	y5	
0.25	0	0.5	1.5	1.589844	1.638153	1.833988	1.863381	1.681916	0.92048705
0.188324	0	0.5	1.5	1.568405	1.604568	1.753716	1.775389	1.638313	0.80850123
0.25	0.188324	0.808501	1.773035	1.856403	1.901019	2.079976	2.10662	1.94126	1.2937217
0.192067	0.188324	0.808501	1.773035	1.83778	1.87192	2.011491	2.031676	1.9036271	17405097
0.20392	0.380391	1.174051	2.029354	2.091427	2.124074	2.255372	2.274263	2.154073	1.61316482
0.206565	0.584311	1.613165	2.271745	2.326045	2.354349	2.465632	2.481366	2.380102	2.10457288
0.214212	0.790876	2.104573	2.479088	2.524275	2.54744	2.634325	2.646171	2.568076 2	6543036
0.225464 1	005088	2.654304	2.644102	2.676656	2.692615	2.74499	2.751256	2.706084	3.26380128
0.24762	1.230552	3.263801	2.749544	2.763568	2.768682	2.767586	2.764947	2.771299	3.94887292
0.25	1.478171	3.948873	2.763882	2.747947	2.735929	2.656062	2.640557	2.722156	4.62774423
0.25	1.728171	4.627744	2.641168	2.586313	2.552099	2.370792	2.338767	2.517156 5	25474897
0.206501	1.728171	4.627744	2.641168	2.596419	2.569447	2.430547	2.406828	2.541928	15141426
0.215663	1.934673	5.151414	2.408456	2.326784	2.278814	2.042876	2.003464	2.231007	5.63074392

Table: Comparison with exact Solution of example above

tk+1	y5	exact	tk+1	y5	exact
0.25	0.92048705	0.920487292	1.230552	3.26380128	3.263801766
0.188324	0.80850123	0.808501278	1.478171	3.94887292	3.948873499
0.438324	1.2937217	1.293721978	1.728171	4.62774423	4.62774482
0.380391	1.17405097	1.174051068	1.978171	5.25474897	5.254749428
0.584311	1.61316482	1.613164996	1.934673	5.15141426	5.151414894
0.790876	2.10457288	2.104573148	2.150336	5.63074392	5.630744518
1.005088	2.6543036	2.654303965			

General Linear Methods

General linear methods (GLMs) for the numerical solution of ordinary differential equations (ODEs)

$$\begin{cases} y'(x) = f(y(x)), \quad x \in [x_0, X], \\ y(x_0) = y_0, \end{cases}$$

are defined by

$$\begin{cases} Y_i = h \sum_{j=1}^{s} a_{ij} f(Y_j) + \sum_{j=1}^{s} u_{ij} y_j^{[n-1]}, \quad i = 1, 2, \ldots, s, \\ Y_i^{[n]} = h \sum_{j=1}^{s} b_{ij} f(Y_j) + \sum_{j=1}^{s} v_{ij} y_j^{[n-1]}, \quad i = 1, 2, \ldots, r, \end{cases}$$

$n = 0, 1, \ldots, N, Nh = X - x_0$. Here, $Y_i, i = 1, 2, \ldots, s$, are internal approximations to $y(x_{n-1} + hc_i), x_{n-1} = x_0 + (n-1)h, y(x)$ is the solution to (1), and $y_i^{[n]}, i = 1, 2, \ldots, r$, are external stages. These methods were introduced by Burrage and Butcher (1980). GLMs include as special cases Runge-Kutta (RK) methods, linear multistep methods (LMMs), e.g. BDF methods, and predictor-corrector methods. Both RK methods LMMs have limitations and the class of GLMs offers new possibilities of constructing new formulas which attempt to combine the advantages of RK methods (large regions of stability) and LMMs (high stage order) at the same time avoiding the disadvantages of these methods (low stage order for RK formulas, small regions of stability for LMMs).

Classification of GLMs

The implementation costs of (2) are determined by the coefficient matrix $A = [a_{ij}]$ and depending on its structure GLMs can be are divided into four types which are appropriate for non-stiff or stiff differential systems in a sequential or parallel computing environment. The construction of GLMs for which the coefficient matrix A takes the following form

$$A = \begin{bmatrix} \lambda & 0 & 0 & \cdots & 0 \\ a_{21} & \lambda & 0 & \cdots & 0 \\ a_{31} & a_{32} & \lambda & \cdots & 0 \\ \vdots & \vdots & \ddots & \ddots & \vdots \\ a_{s1} & a_{s2} & \cdots & a_{s,s-1} & \lambda \end{bmatrix},$$

where $\lambda = 0$ for type 1 methods or $\lambda > 0$ for type 2 methods. These methods are appropriate for non-stiff or stiff differential systems in a sequential computing environment. The type 3 and type 4 methods correspond to the matrix of the form

$$A = \mathrm{diag}(\lambda, \lambda, \ldots, \lambda) = \lambda I,$$

where $\lambda = 0$ for type 3 methods or $\lambda > 0$ for type 4 methods. These formulas are appropriate for non-stiff or stiff differential systems in a parallel computing environment.

Diagonally Implicit Multistage Integration Methods

In the remainder of the paper we will restrict our attention mainly to the methods such that $p = q = r = s$, where p is the order, q is the stage order, r is the number of external stages and s is the number of internal stages. We will also assume that $U = I$, where I is the identity matrix of dimension r. Moreover, we will assume that V is a rank one matrix of the form

$$V = \begin{bmatrix} v_1 & v_2 & \cdots & v_s \\ v_1 & v_2 & \cdots & v_s \\ \vdots & \vdots & \ddots & \vdots \\ v_1 & v_2 & \cdots & v_s \end{bmatrix},$$

with its only nonzero eigenvalue equal to one. This is equivalent to the condition $\sum_{i=1}^{r} v_i = 1$, and as the result this matrix is power bounded which ensures zero-stability of (2). The resulting methods are then called diagonally implicit multistage integration methods (DIMSIMs). This class of GLMs was introduced by Butcher. We believe that the algorithms based on these methods have great potential for practical use.

Order and Stage Order Conditions

It was proved that the method (2) has order p and stage order q=p if and only if

$$B = B_0 - AB_1 - VB_2 + VA,$$

Where

$$B_0 = \left[\int_0^{1+c_i} \phi_j(x)dx/\phi_j(c_j) \right]_{i,j=1}^{s},$$

$$B_1 = \left[\phi_j(1+c_i)/\phi_j(c_j) \right]_{i,j=1}^{s},$$

$$B_2 = \left[\int_0^{c_i} \phi_j(x)dx/\phi_j(c_j) \right]_{i,j=1}^{s},$$

and

$$\phi_j(x) = \prod_{k=1,k\neq j}^{s} (x - c_k).$$

The matrices B_0, B_1 and B_2 can be precomputed for specific choices of the abscissa vector $c = [c_1,\ldots,c_s]^T$ and the relation plays an important role in the construction of DIMSIMs.

Absolute Stability Properties of GLMs

Applying the GLM (2) to the test equation $y' = \zeta y$, where ζ is a complex parameter, we obtain

$$y^{[n]} = M(z)y^{[n-1]},$$

with $z = h\zeta$ and the matrix M(z) defined by

$$M(z) = V + zB(I - zA)^{-1}U.$$

The matrix M(z) defined by (4) will be referred to as the stability matrix of (2) and the rational function defined by

$$p(w, z) = \det(wI - M(z))$$

as the stability function of the method (2). This function determines the absolute stability properties of the general linear method (2). The region of absolute stability is defined as the set where M(z) is power-bounded, i.e. $\{z \in \mathbb{C} : p(w, z) = 0 \Rightarrow |w| \leq 1\}$. A general linear method is called A-stable if its region of absolute stability contains the left half of the complex plane.

Examples of DIMSIMs

Type 1 method with $p = q = r = s = 2$ and $c = [0,1]^T$:

$$A = \begin{bmatrix} 0 & 0 \\ 2 & 0 \end{bmatrix}, \quad U = \begin{bmatrix} 1 & 0 \\ 0 & 1 \end{bmatrix}, \quad B = \begin{bmatrix} \dfrac{5}{4} & \dfrac{1}{4} \\ \dfrac{3}{4} & -\dfrac{1}{4} \end{bmatrix}, \quad V = \begin{bmatrix} \dfrac{1}{2} & \dfrac{1}{2} \\ \dfrac{1}{2} & \dfrac{1}{2} \end{bmatrix}.$$

This method has the same region of absolute stability as an explicit two-stage Runge-Kutta method of order two.

Type 2 method with $p = q = r = s = 2$ and $c = [0,1]^T$:

$$A = \begin{bmatrix} \dfrac{2-\sqrt{2}}{2} & \\ \dfrac{6+2\sqrt{2}}{7} & \dfrac{2-\sqrt{2}}{2} \end{bmatrix}, \quad U = \begin{bmatrix} 1 & 0 \\ 0 & 1 \end{bmatrix}, \quad B = \begin{bmatrix} \dfrac{73-34\sqrt{2}}{28} & \dfrac{-5+4\sqrt{2}}{4} \\ \dfrac{87-48\sqrt{2}}{28} & \dfrac{-45+34\sqrt{2}}{28} \end{bmatrix}, \quad V = \begin{bmatrix} \dfrac{3-\sqrt{2}}{2} & \dfrac{-1+\sqrt{2}}{2} \\ \dfrac{3-\sqrt{2}}{2} & \dfrac{-1+\sqrt{2}}{2} \end{bmatrix} .:$$

This method is L-stable. i.e. it is A-stable and $\lim_{z \to \infty} \lambda_{max}(M(z)) = 0$.

Type 3 method with $p = q = r = s = 2$ and $c = [0,1]^T$,

$$A = \begin{bmatrix} 0 & 0 \\ 0 & 0 \end{bmatrix}, \quad U = \begin{bmatrix} 1 & 0 \\ 0 & 1 \end{bmatrix}, \quad B = \begin{bmatrix} -\dfrac{3}{8} & -\dfrac{3}{8} \\ -\dfrac{7}{8} & \dfrac{9}{8} \end{bmatrix}, \quad V = \begin{bmatrix} -\dfrac{3}{4} & -\dfrac{7}{4} \\ -\dfrac{3}{4} & \dfrac{7}{4} \end{bmatrix} .:$$

This method has the interval of absolute stability equal to $[-\dfrac{4}{3}, 0]$.

Type 4 method with $p = q = r = s = 2$ and $c = [0,1]^T$, Butcher (1993a):

$$A = \begin{bmatrix} \dfrac{3-\sqrt{2}}{2} & 0 \\ 0 & \dfrac{3-\sqrt{3}}{2} \end{bmatrix}, \quad U = \begin{bmatrix} 1 & 0 \\ 0 & 1 \end{bmatrix}, \quad B = \begin{bmatrix} \dfrac{18-11\sqrt{3}}{4} & \dfrac{-12+7\sqrt{3}}{4} \\ \dfrac{22-13\sqrt{3}}{4} & \dfrac{-12+9\sqrt{3}}{4} \end{bmatrix}, \quad V = \begin{bmatrix} \dfrac{3-\sqrt{2}}{2} & \dfrac{-1+2\sqrt{3}}{2} \\ \dfrac{3-\sqrt{2}}{2} & \dfrac{-1+2\sqrt{3}}{2} \end{bmatrix} .:$$

This method is L-stable.

GLMs with Inherent Runge-Kutta Stability

A closely related class of GLMs with so called inherent Runge-Kutta stability (IRKS) was introduced recently by Butcher and Wright. These methods have the property that the stability function p(w,z) defined by (5) takes the form

$$p(w, z) = w^p (w - R(z)),$$

where R(z) is the stability function of a Runge-Kutta method of order p . In contrast to DIMSIMs these methods with $p = q$ and $r = s = p + 1$ can be constructed using only linear operations.

References

- Christian Grossmann; Hans-G. Roos; Martin Stynes (2007). Numerical Treatment of Partial Differential Equations. Springer Science & Business Media. p. 23. ISBN 978-3-540-71584-9

- Eulers Method: tutorial.math.lamar.edu, Retrieved 26 June 2018

- Jaluria Y; Atluri S (1994). "Computational heat transfer". Computational Mechanics. 14: 385–386. doi:10.1007/BF00377593

- Lakoba, Taras I. (2012), Simple Euler method and its modifications (PDF) (Lecture notes for MATH334), University of Vermont, Retrieved 29 February 2012

- General-linear-methods: scholarpedia.org, Retrieved 18 April 2018

- Arieh Iserles (2008). A first course in the numerical analysis of differential equations. Cambridge University Press. p. 23. ISBN 9780521734905

Permissions

All chapters in this book are published with permission under the Creative Commons Attribution Share Alike License or equivalent. Every chapter published in this book has been scrutinized by our experts. Their significance has been extensively debated. The topics covered herein carry significant information for a comprehensive understanding. They may even be implemented as practical applications or may be referred to as a beginning point for further studies.

We would like to thank the editorial team for lending their expertise to make the book truly unique. They have played a crucial role in the development of this book. Without their invaluable contributions this book wouldn't have been possible. They have made vital efforts to compile up to date information on the varied aspects of this subject to make this book a valuable addition to the collection of many professionals and students.

This book was conceptualized with the vision of imparting up-to-date and integrated information in this field. To ensure the same, a matchless editorial board was set up. Every individual on the board went through rigorous rounds of assessment to prove their worth. After which they invested a large part of their time researching and compiling the most relevant data for our readers.

The editorial board has been involved in producing this book since its inception. They have spent rigorous hours researching and exploring the diverse topics which have resulted in the successful publishing of this book. They have passed on their knowledge of decades through this book. To expedite this challenging task, the publisher supported the team at every step. A small team of assistant editors was also appointed to further simplify the editing procedure and attain best results for the readers.

Apart from the editorial board, the designing team has also invested a significant amount of their time in understanding the subject and creating the most relevant covers. They scrutinized every image to scout for the most suitable representation of the subject and create an appropriate cover for the book.

The publishing team has been an ardent support to the editorial, designing and production team. Their endless efforts to recruit the best for this project, has resulted in the accomplishment of this book. They are a veteran in the field of academics and their pool of knowledge is as vast as their experience in printing. Their expertise and guidance has proved useful at every step. Their uncompromising quality standards have made this book an exceptional effort. Their encouragement from time to time has been an inspiration for everyone.

The publisher and the editorial board hope that this book will prove to be a valuable piece of knowledge for students, practitioners and scholars across the globe.

Index

www.ingramcontent.com/pod-product-compliance
Lightning Source LLC
Chambersburg PA
CBHW061958190326
41458CB00009B/2901